More Praise for **Strata**

"Like the earth itself, *Strata* is a work of many layers. It's about the deep past, about how geologists work and think, about the great changes that have taken place in geological history and the ones that lie ahead. Laura Poppick is an elegant writer and an intrepid reporter."
—Elizabeth Kolbert, Pulitzer Prize–winning author of *The Sixth Extinction*

"In the tradition of our best natural history writers, Laura Poppick understands the true gift of geology is the perspective of deep time, where we come to understand that not even stone is indelible. *Strata* offers a reminder that the things which connect us, and will outlast us all, are deeper still: the iron in our blood, the oxygen we breathe, and the stories we tell."
—Rebecca Boyle, author of *Our Moon*

"Laura Poppick paints a fascinating word picture of dramatic change, the hallmark of billions of years of Earth's history—all written in the layers, thick and thin, of the rock beneath our feet. Read her words and know the stories she tells provide solace that change has always been part of Earth's DNA—but rarely at the breakneck speed at which people are altering the planet today."
—Paul Bierman, geologist and author of *When the Ice Is Gone*

"Laura Poppick takes readers deep into the minds of geologists working to interpret the sedimentary chronicles of four critical 'moments' in Earth's past. Revealing the logic used to decode the rock record—and the reasons geologists sometimes disagree about the details of the translation—*Strata* is an extraordinary book."
—Marcia Bjornerud, author of *Turning to Stone*

"In prose as graceful and clear as a mountain stream, Laura Poppick guides us on a journey through the stone palimpsest that is Earth's crust, revealing the hidden histories upon which we walk every day. Poppick weaves scientific exposition with well-chosen anecdotes, portraits of contemporary scientists, field reporting, and strands of memoir and poetry, forming a tapestry as splendid as Earth's richly layered skin. . . . An excellent choice for anyone who wants to better understand not only how our planet came to be the world we inhabit today, but also why we must continually relearn how best to interpret the vestiges of its unfathomably vast, bewilderingly complex, and endlessly fascinating past."
—**Ferris Jabr,** author of *Becoming Earth*

"Rock has never felt more alive, nor deep time more current, than in Laura Poppick's absorbing, illuminating *Strata*. In the intrepid tradition of John McPhee and Elizabeth Kolbert, Poppick spelunks into our planet's history to unearth the ancient dramas that sculpted our landscapes and ourselves. This book is an indispensable guide to the dynamic stories our planet writes in stone."
—**Ben Goldfarb,** award-winning author of *Crossings* and *Eager*

Strata
Stories from Deep Time

Laura Poppick
Illustrations by Sarah Gilman

W. W. Norton & Company
Independent Publishers Since 1923

Copyright © 2025 by Laura Poppick

All rights reserved
Printed in the United States of America
First Edition

Selected material in this book originally appeared in *Knowable Magazine* from Annual Reviews as "The Story of Snowball Earth" (2019), "To Date a Dinosaur" (2019), and "The Origin of Mud" (2020).

Selected material in chapter 13 originally appeared in "From Ancient Charcoal, Hints of Wildfires to Come" by Laura Poppick © 2024 The New York Times Company.

For information about permission to reproduce selections from this book, write to Permissions, W. W. Norton & Company, Inc., 500 Fifth Avenue, New York, NY 10110

For information about special discounts for bulk purchases, please contact W. W. Norton Special Sales at specialsales@wwnorton.com or 800-233-4830

Manufacturing by Lakeside Book Company
Book design by Lovedog Studio
Production manager: Devon Zahn

ISBN: 978-1-324-02160-5

W. W. Norton & Company, Inc.
500 Fifth Avenue, New York, NY 10110
www.wwnorton.com

W. W. Norton & Company Ltd.
15 Carlisle Street, London W1D 3BS

10 9 8 7 6 5 4 3 2 1

For my family.

The sediments are a sort of epic poem of the earth. When we are wise enough, perhaps we can read in them all of past history.

—RACHEL CARSON, *THE SEA AROUND US*

Contents

Prologue   1

Part I
AIR   9

Part II
ICE   73

Part III
MUD   133

Part IV
HEAT   185

Epilogue: Us   235

Acknowledgments   243

Notes   245

Index   268

Heat
252 million years ago
Mesozoic hothouse

Air
2.4 billion years ago
Rise of oxygen

Mud
458 million years ago
Rise of mud on land

Ice
720 million years ago
Snowball Earth

Formation of Earth
4.54 billion years ago

Prologue

Consider a single grain of sediment.

A piece of sand knocking along a riverbed. Silty dust whipped off a desert dune. The flake of a broken shell, feathering to the bottom of the sea.

Earth is in a constant state of shedding, releasing bits and pieces that sift and gather in low places and accumulate as layers over time. If left undisturbed, these layers turn to stone, fossilizing moments from Earth's deep past.

When these moments later resurface as mountains or cliff faces, we can put our fingers to those strata, read them like lines in a poem. We can find scenes of ancient storms and mass extinctions, droughts and evolutions, traces of perpetual and inevitable change.

We can learn how we arrived here, in this moment, now.

✦ ✦ ✦ ✦

THE WALLS OF WYOMING'S Bighorn Canyon rise above a slender lake that once ran as a river. That river flowed for many thousands of years, surging at the end of the last ice age as melting glaciers sent torrents down the Rocky Mountains and out into tributaries that carved through the earth like blades through soft lumber.

That flowing and carving ended here, in the 1960s, with the construction of a dam. Ever since, the waters have lapped calmly against sedimentary layers that ascend more than a thousand feet, capped by a plateau on top.

A web of hiking trails stretches across that plateau today in the Bighorn Canyon National Recreation Area, where I found myself walking a few summers ago. As I crunched along a dusty path, I dutifully obeyed signs warning me to watch out for scorpions and rattlesnakes. I peered behind bushes and beneath boulders for a mile or so, until I followed a bend in the path and came upon that massive scar in the earth, the Bighorn Canyon.

I set down my backpack and folded my legs beneath me to sit on the pale ground. Bluebirds chased flies through juniper and sagebrush; wild mustangs grazed on a hillside nearby. I could hear families shouting about snakes or scorpions up ahead. But all that faded around me as I took in the strata, those neatly arranged layers of sand, silt, and clay laid out on the canyon walls in a striped pattern, all the way down to the water below. Pulled from the moment, I plunged into deep time, caught somewhere in the Paleozoic—long before human voices echoed on Earth, even before snakes or hooves or grasses grew here.

I sat for a while, taking in the remnants of ancient beaches or seafloors or other sandy environments that had long since turned to stone. My mind shuffled through the winds and rains that had spilled out those sediments, and all the dramas that had unfolded in the millions of years since then. The rise and fall of species; the coalescing and disintegration of continents. The storms bigger than the last, the clouds too heavy to hold snow.

I imagined my own short life in the context of all the time locked up in that canyon wall and felt tiny. Tinier than tiny, a speck of dust.

In that smallness, I felt a sort of relief.

On my way back to the trailhead, I stumbled over a patch of crumbly rock, an extension of the canyon wall. I crouched and found a collection of perfectly sculpted ripples, ones you might find at the edge of a sea. I traced the ridges with my forefinger and felt curves shaped by flowing water millions of years earlier. A moment frozen in time, whispering memories of a world without us.

PROLOGUE

✦ ✦ ✦ ✦

SO MUCH OF WHAT we know about Earth's 4.54-billion-year history, we know thanks to strata like those in that canyon wall in Wyoming. It is through these lines in stone that we can glimpse ancient iterations of this planet and gain context for the moment we're spinning through now. Along canyon walls and hillsides on all seven continents, geologists flock to these types of sedimentary rocks to piece together the story of how Earth became the place we inhabit today. They collect bits of stone that they bring back to the lab to puzzle out how our atmosphere formed into the air we breathe, or how the planet's thermostat has fluctuated through time, or how landscapes have morphed, and how life has shaped and been shaped by all these upheavals. They turn to these rocks to understand the underpinnings of our current environmental crises, searching for evidence of past periods of tumult to help get us through this one.

But the language of stratigraphy is not in the common lexicon. Most hikers along that canyon wall in Wyoming probably take in the wide sky and the river down below, and then carry on, warily scanning for venomous things without giving the strata much thought. I had arrived there with an awareness that bordered on reverence for those lines in stone, but only because of a geology class I had stumbled into during my freshman year in college. Before then, I wouldn't have given much thought to the strata either.

In that first geology class, we traveled to lakes and rivers and beaches all across Maine, collecting samples of silt and sand that we brought back to the lab and dropped through sieves and weighed on scales to tease out patterns in the landscapes around us. I hadn't signed up for that class with any particular interest in sediments. I mostly just wanted an excuse to tramp around outside for credit. But over that semester, I grew to appreciate the silt and sand for the ways they drew my attention to details in the landscape I had never thought to pay attention to before. How the inner bend of a river flows more slowly and drops finer sediment than the outer bend, or how one side of a beach may contain larger sediments than the opposite side. How this sorting isn't random, but is orchestrated by winds and tides and currents that are, in turn, orchestrated by a larger web of activity I came to know as the Earth system. This hidden organizing force, I learned, extended all around us, weaving a precise order through even the smallest of grains.

I declared myself a geology major and, over the next four years, began to see the world through a wholly different lens. Landforms I had once perceived as blank backdrops for the travails of daily life began brimming with their own quiet sagas. I found stories in the curves in the land around campus, couldn't jog through town without thinking of the glaciers that had carved the river I ran beside. I stopped to examine the foundations of buildings and the contents of slate walkways, often to perplexed looks from my friends.

I went on to work as a field assistant and lab technician for a group of stratigraphers for a year or so after I graduated, before replacing my rock hammer with a voice recorder and reporter's notebook. Still digging into stories of the planet, but from the human perspective rather than that of the rocks. Even after I left geology for journalism, though, my appreciation for strata only grew. The crises facing our planet increasingly called for the insights and context tied up in those layers, for the wisdom that they held.

By the time I arrived on that trail in Wyoming, I had been working as an environmental journalist for more than a decade, covering story after story of Earth's unraveling. I had spoken with glaciologists about disappearing ice, oceanographers about disappearing coastlines, geochemists about disappearing permafrost. I arrived at that canyon wall with a longing for a version of the planet I knew I would never see in my lifetime. A version that wasn't burning, flooding, skittering off-kilter from our own undoings.

As I sat on that pale plateau with my legs beneath me, though, I remembered that stability has come and gone and returned so many

PROLOGUE

times before now. That geologic timescales arc too wide to witness in a single human lifetime, but have always spun toward some sort of new stasis. I knew this didn't let us off the hook, or mean that it was time to stop righting our wrongs to the environment. The changes we have unleashed today are unfolding far faster than past periods of change, and they were not geologically inevitable. We are the agents of this geologic moment. But the strata reminded me that we are also part of the Earth system, this much larger web of connections that thread between the atmosphere, continents, water, ice, and life. That these threads slacken and tighten over time and accommodate for one another with more brilliance than the human mind can easily grasp. That we live within this system, and the system lives within us. We carry its iron in our blood and its stardust in our bones, and its strength is our strength because we are it.

We are it, but we are not a given. The only given is the change and the sphere that contains it.

✦ ✦ ✦ ✦

WE HAVE FOUND OURSELVES at a turning point, not only in our global environmental crises, but in our capacity to do something about them. Never in human history have we known so much about what came before us, and never have we been so equipped to do something with that knowledge and reroute our path forward, so long as those in power are willing to do so.

Beyond an accrual of knowledge, I have found that my own understanding of strata has given me a deeper and still deepening love of Earth in all its layered complexity. To know a history is to grow in intimacy with a place, and this has been geology's gift to me.

This is what I want for you in reading this book. I want you to bask in this current moment, in the awe that we get to be here at all.

I also want you to know that, millions of years from now, our time on Earth will appear as its own hardened layer in cliffsides and canyon walls. A future being may come across that layer, trace it with their toe, and try to make sense of their own turmoil by decoding the details of ours.

Beneath that layer, they may find traces of stories far older than ours. Some of them might be familiar to you: the rise and fall of dinosaurs; the most recent ice age. But others will be less familiar, and I've chosen a selection of these less familiar tales to tell in the pages ahead. Each illustrates a moment of global transformation that incrementally shaped the planet as we know it today and made our lives here possible. They mark periods when Earth fell into some sort of disarray, then bounced back to a new stasis. They demonstrate how the threads of the Earth system—the atmosphere, water, rock, ice, and life—interweave to prop up the planet through periods of turmoil. They show us how we found ourselves in our modern crises, and how we might find a way out of them.

By leapfrogging across geologic time, we'll see how Earth didn't appear fully formed waiting for us to arrive, but instead spun through a series of intense growth spurts that cumulatively, over billions of years, built the landscapes we now have the privilege of inhabiting. We'll see how a bacterial blob evolved to turn sunshine into sugar and filled the atmosphere with oxygen for the first time; how a series of global ice ages may have paved the way for multicellular life; how the rise of plants on land instigated an onslaught of mud that rewrote the surface of the planet; and how past periods of heat can tell us something about our moment of warmth today.

These through lines rest beneath our feet, offering wisdom that we may carry into our own lives, so long as we know how to read them.

Let me show you how.

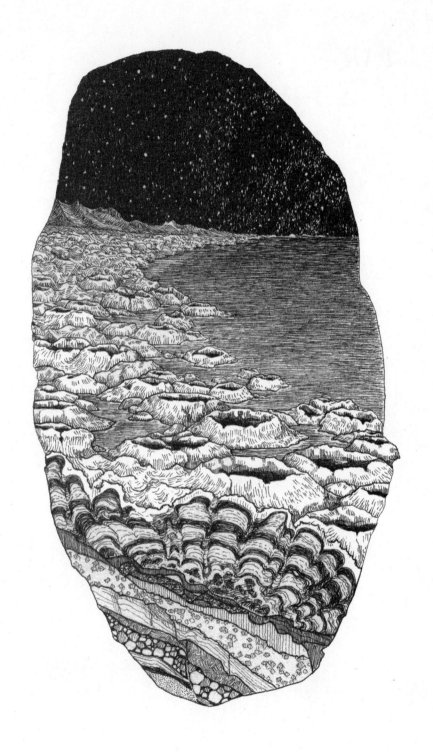

PART I
AIR

To read the planet's earliest strata is to know a time older than bones. A time before lungs or lips, seeds or whiskers. When no shells jostled in the incoming and outgoing tides, and no seaweed rotted along these shores.

A time before fire, when there was nothing to burn.

Some elements would have been familiar to us. Rocks, clouds, sun, and sea. Wind lapping water against lake shores. But life and all its riots would have been limited.

Then, just as the atmosphere of a room shifts as new bodies enter, so too did the atmosphere of Earth change as new lifeforms emerged.

Most traces of these earliest beings have long been lost to the mantle, their strata melted deep beneath granite crust and spewed back out as gases and minerals that have since precipitated lungs, lips, bones, and moods.

We inhale their remnants in the air we move through.

One

DOWN THE STREET FROM AN APARTMENT I USED TO rent outside of Portland, Maine, sits a beach scattered with broken bric-a-brac. Depending on the time of day you visit, calling this strip of land a "beach" is actually a bit of a misnomer. At high tide, it vanishes beneath the cold waters of Portland Harbor. At low tide, it morphs into a vast mudflat that slurps off the shoes of anyone willing to tread to its edge. Only at mid-tide does it really look and behave like a beach, with a sandy strip that runs about a quarter mile in length.

The first winter I lived there, I developed a habit of visiting this strip of sand every day. When the tide allowed it, I walked the entire length a few times, adding texture to weeks when time otherwise bled together. I began those walks with a nod to the waves before shifting my gaze downward to see what the tide had draped out since the day before. Neon scraps of fishing line. Rusted scrap metal. Lobster traps. Seemingly endless fragments of glass.

The bits of detritus that intrigued me most, though, were the shards of porcelain dishware that had broken who knows how long ago. Decades? Centuries? The mystery of their history drew me in. On any given walk, I would find at least a dozen of these porcelain pieces. Many were plain white and not, in my mind, worth pocketing. But if I squatted to take a closer look or turn them over, I would sometimes discover the cobalt outline of a rose, or the frilly ochre edges of a leafy design.

As the winter wore on and my porcelain collection grew

from a handful to a jarful, I wove stories of how I imagined those ornate shards came to rest there. They were the remnants of a shipwreck; they were thrown overboard to lighten a load; they were the result of a Victorian woman smashing her dishware against the rocks to blow off some steam. (This last one was my preferred version of the story.)

No matter how those fragments got there, their presence alone revealed some plain truths about that beach. They spoke of the continent's first porcelain-bearing humans, and the displacement of the citizens of the Wabanaki Nations who had lived on and fished those shores for millennia before European settlers arrived. The shards held remnants of not only a single event—the shattering of a dish—but infinite other moments of loss and change.

Then we have the porcelain itself. The quartzes and feldspars that made up the clays that made up the material that made up the dish that broke. Those minerals come laced with their own narratives of making and unmaking and making anew. If we look beyond the surface of debris left behind by our own species, we find mirrors reflecting and receding and reflecting all the way to Earth's beginnings. We can muck our way back to a time before there was any human power to make or land to take. Before there were lungs to breathe out words and stories of where we all came from and where we might be going.

+ + + +

IF YOU WERE TO slice Earth in half and drag your finger from top to center, you would feel three concentric layers: a cold and brittle crust, a warm and gooey mantle, and a hot, dense core.

The stories in the pages that follow take place mostly atop the outermost layer of the planet. As this cold, rocky rind erodes and breaks down over time, it recycles itself into sediments that, in turn, solidify into sedimentary rocks—the fibers from which stratigraphers weave their narratives. These hardened seafloors or lake beds or dune fields

AIR

or other such sediment-rich environments have turned to stone under their own weight, with the help of heat and pressure from the mantle. Unlike igneous and metamorphic rocks, which form deeper within Earth's insides, sedimentary rocks preserve pieces of Earth's ancient surface and the life that whizzed around on top of it. A sedimentary rock may get pushed back toward the mantle and then later pop up as a metasedimentary rock with some of its original narratives still intact. But it's the unadulterated, unmetamorphosed sedimentary deposits that provide the clearest views of the events that unfolded atop Earth's crust.

To the untrained eye, these rocks might appear mundane. Beige, white, gray. Some silt and sand. What of interest could they possibly tell us? But to the trained eye, they contain physical and chemical clues, or proxies, that reveal in remarkable detail how the planet looked and felt at the time the rocks formed. Some of these proxies are visible to the naked eye, while others only emerge through laboratory analyses. Some provide information on a local level, whereas others reveal global phenomena. The shape and height of an ancient dune, for example, might say something about the strength and direction of wind in a particular location, whereas the geochemistry of ancient seafloor sediments might reveal something broader about global seawater temperatures at that time. While pieces of broken porcelain on a beach might provide proxies for relatively recent human activity, these geologic proxies open portals to deep time. It's by piecing together these physical and chemical proxies within sedimentary rocks that stratigraphers have sharpened our understanding of how this planet looked and felt long before we arrived.

The problem with proxies, though, is that they are not always what they seem. Red herrings abound. One look at today's world reveals how easy it can be to misinterpret sedimentary features on the landscape. A smattering of circular indentations on a sandy beach, for example, might seem to be a proxy for rain. The dots are not the raindrops themselves, but stand-ins for the water, an indication that something physical once fell from above. I can walk a beach hours after a storm and confidently conclude how those indentations formed. But the round, shallow depression in the sand that I think came from rain could actually have come from, say, the tip of a walking stick. Or a toddler dipping their finger in the sand. Or a hundred other random occurrences that may have taken place while it was also, in fact, raining.

The stratigrapher must always remain wary.

Sedimentary layers grow more difficult to read as they grow older in age and become overwritten with new narratives. As the rock record goes back in time, writes John McPhee, "it touches upon tens of hundreds of stories, wherein the face of the earth often changed, changed utterly, and changed again, like the face of a crackling fire."

Trying to get a handle on Earth's oldest stories is like trying to contain a whipping flame. All but impossible. Strata from the very earliest eon, the Hadean, have all but melted back into the mantle. We'll pass over these very earliest days and make our way into Earth's second eon, the Archean. Spanning from 4 billion to 2.5 billion years ago, this marks the beginning of the rock record, and the beginnings of the air that we breathe.

✦ ✦ ✦ ✦

AS YOU READ THIS LINE, the oxygen you are pulling inside your body makes your body possible. It is allowing you to digest your most recent meal, move your eyes across these words, and think your thoughts. It is the single most important gas to your survival. You share this in common with every other animal on Earth, save for one lone parasite of Chinook salmon that somehow doesn't need it. Well done, *Henneguya salminicola*.

Throughout a given day, you fill your lungs with oxygen some 20,000 times. Most of us probably don't give it much thought. Maybe we assume that this gas has always been here, a given on this highly habitable pale blue dot.

But it turns out that this dot has not always been highly habitable, nor, for that matter, has it always been blue. The early Earth's young magma surface sat gooey and cloaked in steam, too hot to hold liquid seas. It took a long time for continents to rise up and for ocean basins to fill in, and far longer still for oxygen to pool up in the atmosphere.

"And so," writes Rachel Carson in *The Sea Around Us*, "the rough outlines of the continents and the empty ocean basins were sculptured out of the surface of the earth in darkness, in a Stygian world of heated rock and swirling clouds and gloom."

Even in those earliest of gloomy days, oxygen—the element O—was all over the place, bound up in molecules like water vapor and quartz and carbon dioxide. It's the third most abundant element in the universe, and it has been present on Earth since the beginning. But free oxygen—two atoms of O bound together by a pair of shared electrons, liberated from any other material but itself—didn't emerge as a gas until more than halfway through Earth's existence.

If you reach out your arms and imagine Earth's 4.54-billion-year history as a timeline that extends from the tip of your right hand to the tip of your left, the arrival of oxygen gas falls around your heart, at about 2.4 billion years ago, give or take a couple hundred million years. (From here on out, when I list a geologic age, know that there is always going to be a healthy margin of error. And when I say oxygen, I mean this gaseous form.)

The fashionably late arrival of oxygen may sound like a planetary sigh of relief. Finally, the possibility for life larger than one cell, with lungs and lips and all the rest of it. But scientists familiar with oxygen's highly reactive habits suggest its arrival was more like a nightmare.

As is true of all elements, an atom of oxygen contains a cloud of negatively charged electrons that spin in an arrangement of "shells" around a positively charged nucleus. The outermost electron shell constantly seeks stability by filling to its capacity. In oxygen's case, its outermost shell is two electrons short—comparatively fewer than other elements—and the configuration of those electrons contributes to oxygen's high reactivity. Oxygen's electron cloud is also relatively thin compared to other elements. Without much of a barrier between it and the outside world, the positive pull of the nucleus easily seeps out and lures in the negative charges of the two electrons it needs to stabilize. Two atoms of oxygen bound together as oxygen gas have a pull similarly as strong as a single atom on its own.

When oxygen first appeared on Earth, it desperately rooted out and bonded with anything willing to share its electrons, fundamentally changing the materials it bonded with. It weaseled into microbial cells and mutilated their machinery. It sulked into currents and eddies and made arsenic more soluble, it spread hydrogen peroxide poisons into DNA. With all the havoc it wreaked, this gas might have initiated one of the worst mass extinctions in all of Earth history—though it's hard to know this for sure, since the single-celled beings that would have gone extinct were too squishy to leave behind reliable fossils. Even so, some call this geologic moment the Oxygen Catastrophe.

Over time, molecules from the bottom of the ocean to the top of the atmosphere grew to accept oxygen's reactivity, and living things evolved ways to cope with this new gas. Their cells grew to tolerate it, and then to depend on it. They used it to break down food and generate energy that allowed them to grow larger and more complex, with multiple cells that communicated across newly sophisticated membranes. These oxygen-fueled innovations expanded and cascaded and eventually led to the evolution of eyeballs and brains and lungs and lips and, over billions of years, the possibility of us.

So what, exactly, happened around 2.4 billion years ago? Why did oxygen arrive when it did? And how can we read this in the rock record?

✦　✦　✦

THE SEARCH FOR OXYGEN'S origin began with a problem.

When Charles Darwin published *On the Origin of Species* in 1859, he agonized over the seeming absence of fossils in the planet's oldest rocks. The ages of rocks at this time were known only in a relative sense—as in, what formed first and what followed. The scientific law of superposition, proposed by Danish geologist Nicolas Steno in the seventeenth century, helped clarify that younger strata always sit atop older strata, since that's how sediments accumulate in lake beds and seafloors and so on.

As hard as paleontologists of that time looked, they couldn't find any remnants of ancient life in the oldest, bottommost strata that they examined. Then bits and bobs appeared in what looked like an explosion of living things in strata above a certain age. This troubled Darwin deeply. Any such explosion of life undermined his theory of natural selection, a process of elimination that he argued should inherently take a very long time to unfold. By his estimations, it could never have taken place as instantaneously as those earliest fossils suggested.

Halfway through *On the Origin of Species*, he gravely acknowledged the implications of this predicament. "The case at present must remain inexplicable[,]" he wrote, "and may be truly urged as a valid argument against the views here entertained."

But here we are, still entertaining Darwin's views more than 150 years later. And that is thanks largely to rocks discovered not long after World War II.

At the end of the war, a wave of mineral exploration arose across the world to meet the needs of rapidly expanding economies. Federal agencies hired geologists to scour continents for oil, gas, and coal to fuel those economies, along with metals like iron and uranium to build up arsenals of defense. This was of national interest, not just private economic interest.

As geologists marched around the globe and sketched up their maps of these resources, they noticed other curious details about the planet's history. That is, in their search for the materials that humans desired, they found inklings of how we got to be here desiring anything in the first place.

In the summer of 1953, Wisconsin geologist Stanley Tyler was studying iron-rich rocks on the north shore of Lake Superior in Canada when he took a Sunday off to rent a boat and go fishing. While his lure bobbed in the water, he absently noted the shapes and colors along the shore, as any geologist might. One outcrop caught his eye, so he motored over to take a closer look.

Tyler recognized the deposit as an extension of the Gunflint Chert, a rock formation with the texture of tightly packed brown sugar and the contents of ancient seafloor sediments. Cherts can take on a whole range of colors depending on the conditions they form within, from beige to red to green to other hues in between. Most of the chert Tyler had found on that trip had been maroon, but this outcrop caught his eye for its striking shade of jet black. He knew that the color black in rocks was sometimes indicative of organic material, remnants of ancient life.

He lopped off a chunk, stashed it in his boat, and motored on.

Back at his lab in Madison he placed a sliver of that black chert under a microscope, and found shapes that did not speak the language of minerals. The rods, spheres, and squiggles he found did, as he suspected, look more lifelike than lithic.

Based on geologic maps of the region, he knew these rocks had formed during the allegedly fossil-free epoch that had so troubled Darwin. Tyler's gut told him he may have just found some of the earliest evidence of life ever discovered, but he was a mineralogist more than a paleontologist and so he needed a second opinion.

That fall, he took photographs of his findings to a geology conference in Boston and shared them with a couple of colleagues. One among them, a Harvard paleobotanist named Elso Barghoorn, agreed that the samples looked rather lifelike, and the two published a short paper describing what they had found.

This publication quadrupled the length of the fossil record. It was groundbreaking, but was brief and preliminary. They needed more time to study the fossils to do justice to the scope of their findings.

For years, they didn't make progress on a follow-up paper. A decade went by and, in 1963, Tyler passed away at the age of fifty-seven from heart complications, without the satisfaction of sharing his discoveries more completely with the world. By 1965, an impatient colleague named Preston Cloud—a bantamweight boxing champion turned acclaimed Earth historian—threatened to beat Barghoorn to the punch with his own paper on the fossils. That was enough to push Barghoorn into gear. He rushed to complete a manuscript and published it in the journal *Science* a couple months before Cloud published his.

"For all of time it will probably stand as the most important article ever written in the field," writes William Schopf, a graduate student who helped Barghoorn pen that manuscript, but who humbly declined authorship himself because he didn't feel he had contributed enough.

Spurred by this new paper on the Gunflint Chert, geologists went searching for evidence of ancient life in black cherts around the world. Papers flooded out, claiming to have solved Darwin's dilemma and showing how fossils *had* been in those seemingly lifeless rocks all along—they had simply been microscopic. The theory of natural selection persevered, and the lengthy record of our ancient roots began to fill out.

But while those microscopic rods and squiggles resolved one nagging dilemma, they opened up a slew of other questions. What, exactly, were those fossils? What kind of world did they evolve into? And what kind of world did they create with their growth?

Around the same time that these questions began bubbling up, another set of observations from the rock record thickened the plot of the squiggles. Geologists were compiling evidence that, before those lifeforms lived, the planet's atmosphere had no oxygen gas in it at all. Minerals that disintegrate in the presence of oxygen were found locked up in ancient riverbeds older than a certain age. Then, around the time they believed those squiggles showed up on the scene, those riverbed minerals disappeared and the very first, rusty red fingerprints of oxygen began appearing in strata around the world.

Perhaps, some reasoned, those squiggles were responsible for painting the world's soil and seafloor sediments red, by ushering in the very first poofs of oxygen. And perhaps, in their delivery of this gas, they catapulted Earth out of its original barrenness and into the tangle of complex life we know today.

✦ ✦ ✦ ✦

EVEN BEFORE EVIDENCE FOR oxygen's initial absence on Earth began cropping up in the rock record, scientists had long thought that the planet's earliest atmosphere likely lacked oxygen. But they based this assumption more on theory than on physical evidence. The theory went that Earth must have made its own atmosphere (rather than taken it from outer space), because the contents of the modern air looked nothing like the contents of the cosmos. Neon, xenon, krypton, and other obscure gases that are exceedingly rare here are common out there. Earth's original air must have come, instead, from inside the planet. Gases billow out of Earth by way of volcanoes, so volcanic emissions must have been responsible for that original atmosphere. Since volcanoes do not emit oxygen in any significant volumes today, they presumably never have, and so the earliest air must have been free of this gas.

The scientists trying to work out this theory—including that bantamweight boxer, Preston Cloud—understood that if oxygen hadn't gotten here by way of volcanoes, then it must have gotten here by way of life. That is, by way of organisms that use the power of the sun to tear apart water molecules to make sugar, and release oxygen as a sort of exhale: things that photosynthesize. A handful of non-biological reactions can create oxygen, but not in quantities that compare with photosynthesis.

If photosynthesizing beings single-handedly oxygenated our atmosphere, as Preston Cloud and others assumed, then the planet's very first photosynthesizers had led the evolutionary bandwagon that would eventually make today's world possible.

Elso Barghoorn thought these earliest oxygen producers must

have been algal, since algae are among the simplest organisms that emit oxygen in today's world, and thus must have sat at the very base of the evolutionary bush of photosynthesis. He identified some of the squiggles in the Gunflint Chert as blue-green algae—a group that lives today and grows thin, sticky mats that trap sands and silt and harden into stone mounds called stromatolites. These unusual knee-high lumps of limestone grow off the coasts of the Bahamas and Western Australia in the modern world, and it was Barghoorn's familiarity with these modern stromatolites that allowed him to correctly identify the ancient ones in the Gunflint Chert.

But while his comparison of ancient and modern stromatolites was spot on, his and his colleagues' assumption that these mounds were algal was not. As paleobotanists, they had looked at the world through the lens of plants, and that lens had led them astray. By the 1970s, new findings indicated that the organisms named blue-green algae were not algal at all, but rather were bacterial. By the end of the century, most researchers had accepted this categorical snafu and had begun calling these organisms by their new name: cyanobacteria, in reference to their blue-green, cyan hue.

The consensus today points to cyanobacteria, or their progenitors, as the earliest oxygen producers on the planet. But recent genomic research has revealed that these microbes didn't whip up their recipe for photosynthesis from scratch. They sit far from the base of the origins of life, which means they most likely evolved from other forms of microbes that used the sun to make food in more basic ways. Instead of breaking apart water to get the electrons that they needed to make sugar, these simpler photosynthesizers made their food by breaking apart fumes of things like iron and sulfur that emanated as gases out of the seafloor. As a byproduct, they released gases of iron and sulfur back out into the environment, rather than oxygen. These types of bacteria continue to live today.

These earlier, ancient microbes would have opted to use seafloor gases in their production of sugar because water is rather difficult to tear apart. It takes an enormous amount of cellular energy to pull those two hydrogens from that one oxygen. Iron and sulfur molecules, on the other hand, are far easier to split apart.

But the simpler form of photosynthesis also had a downside: iron and sulfur gases emanated out of the crust only in discrete locations. These were mere geochemical trickles in a vast ocean. Water, meanwhile, sloshed everywhere. If they could figure out how to muster the energy to turn that liquid into food, then the habitat available to them would explode—as would their populations.

So when, exactly, did bacteria evolve the capacity to use water and sunlight to make sugar and oxygen? How can we read this in the rock record? And why is it worth trying to decode this story today?

Two

BENEATH THE SHADE OF A JACARANDA TREE, I watched a squirrel nose the dry grass beside me. He stopped, dug a hole, dropped something into it, and then caressed the earth back over the stash with careful paws. When finished, he looked left, then right, then straight into my eyes before darting off to dig another hole in the ground beneath the tree.

It was a moment that could easily blend into the backdrop of a day. The grass and the animal and the breeze that tickled my neck were all so familiar, seemingly givens of this world we live in. But I was there in the outskirts of Los Angeles to learn about a time in Earth's past when none of this would have existed. When paws and stems hadn't yet evolved, and shade came only from the rocky relief of hills and mountains.

I met Woody Fischer, a geobiologist at the California Institute of Technology, in that courtyard on an unseasonably cold October morning, where he greeted me with a face-consuming smile and sturdy handshake. He's one of the leading figures working to untangle oxygen's early origins on this planet, and had agreed to walk me through all that we've learned so far about the rise of this gas in the atmosphere.

I had known of Woody for years—and had even coauthored a paper with him during my brief stint working in geology—but this was our first time meeting in person. I was quickly pulled into the orbit of someone who can talk for hours without stopping to rest and whose depth of knowledge peels back as endlessly as an onion skin. His graduate students describe him interchangeably as the most positive person they know and the most talkative person they know, and I scrambled to keep up as he led the way down the succulent-lined sidewalks of Pasadena to a nearby coffee shop to chat.

As a kid growing up in Minneapolis in the 1980s, with a mother who taught ballet and a father who worked as a banker, Woody didn't have his sights set on becoming a scientist. He didn't even think of himself as particularly driven at school, nor did he like it very much. But when he took his first geology class as a freshman at Colorado College, his view of the world shifted. "It makes you feel small," he told me. "And it should. It's working if it does."

This interest in Earth history had, it turns out, run in his family. His father had also studied geology at Colorado College, and his grandfather had taught geology there for decades as well. Accepting this inheritance, Woody declared himself a major. But he still left other doors open. He joined an improv group and formed a '60s surf band called Vladimir Chimp and the Space Barnacles. He considered a career in law.

All the while, though, his interest in geology and the history of the planet continued to grow and, eventually, Woody gave in. In 2002, he enrolled as a PhD student at Harvard in the newly emerging field of geobiology—the study of the cross talk between Earth's surface and the life that inhabits it. Though he may not have considered himself much of a scientist earlier in life, he made a strong impression when he first arrived on campus. "He was amazingly well-read," says Paul Hoffman, a geologist who worked at Harvard at the time. "He felt like an addition to the faculty."

The same year Woody began graduate school, a Harvard geochemist named Dick Holland published a paper on the rise of oxygen and coined this pivotal moment the Great Oxidation Event. (This term is still used today, though some prefer the term Great Oxygenation Event.) Research on this event, called the GOE for short, had been going on for decades but was enjoying a bit of a heyday at that time. It appealed to Woody's interest in moments of global

AIR   25

transformation—when things got difficult on Earth and life noticed, as he puts it, or when life figured something out and Earth noticed.

He dug in and began tackling some of the many lingering questions about the onset of the GOE from his own angles, including how and why it came about when it did, and how we can say this with certainty from evidence in the rock record. He worked under the mentorship of Earth historian Andy Knoll, who himself had studied with Elso Barghoorn in the 1970s. Woody considers Barghoorn to be his academic grandfather.

As he plugged away at his research questions, Woody maintained a fairly simple outlook on the onset of the GOE. The way he and others in his camp saw it, the changes in the rock record that define the GOE must have come about with the rise of oxygen-producing photosynthesizers: cyanobacteria, or their progenitors. Full stop. Getting to the bottom of the GOE would mean getting to the bottom of the evolution of these microbes—how they evolved when they did, and how we can say this with certainty from clues in the strata.

Why go to such lengths to pinpoint the origins of this group of microbes? Cyanobacteria are, to Woody's mind, the greatest environmental engineers in the planet's history. With their production of oxygen, they rewrote the chemical composition of Earth from the seafloor to the stratosphere. Their arrival on Earth has something to teach us not only about our past but also about our future, by demonstrating how the planet responds to major environmental perturbations.

Woody continues to chip away at questions surrounding the rise of oxygen from his lab at Caltech. He takes an interdisciplinary approach, wearing both the geology and biology hats required of geobiology, along with the chemistry hat of geochemistry and the combination of all of these in the catchall field of biogeochemistry. On any given day, he and his students might find themselves puzzling over the chemical composition of an ancient riverbed, developing mathematical models to better understand the underpinnings of geochemical proxies, or clanking around beakers of modern cyanobacteria to better understand how their biology works today. Because Woody alone can't realistically master each of these subfields of Earth history, he makes a point to take on graduate students who collectively can. Some are whizzes at computer modeling, others are talented geochemists, others are excellent field geologists. Together, they are working to triangulate data collected from some of the oldest sedimentary environments to zero in on the instigators of the GOE.

Since his time at Harvard, some details of the GOE have come into sharper focus. The timing of the event has been dialed in around 2.4 billion years ago, give or take a couple hundred million years, thanks to "smoking gun" evidence compiled from changes in the sulfur chemistry of seafloor sediments of this age—changes that could have only come about with the absence and then presence of oxygen. It's also clear that when oxygen arrived, it did not seep across the entire planet all at once, or accumulate very quickly. For hundreds of millions of years, it made up less than 1 percent of modern levels, in an atmosphere otherwise suffused with carbon dioxide, methane, and water vapor. It took something like 2 billion years beyond the GOE to reach concentrations high enough to fully oxygenate the ocean, and from there it roller-coastered up and down before finally arriving at modern concentrations of 21 percent. Why oxygen took so long to reach today's concentrations remains an open and intensely studied question, as does why it first arrived 2.4 billion years ago.

The fact that so many unknowns persist around the GOE after decades of research is not so much a reflection of the quality of the science but of the genuine challenge of pulling certitudes about the ancient atmosphere from some of the planet's oldest rocks. These strata have existed for more than 2 billion years. In that time, they have been pushed, shoved, mangled, and rewritten over and over again. Detangling the narratives and through lines of deep time takes, well, a lot of time.

As the years have spun on, however, a growing body of research has called into question the simplicity of Woody's angle on the GOE. Nobody denies that cyanobacteria or their precursors brought oxygen into the world. That's a closed case. But the extent to which their debut instigated the accumulation of oxygen that marks the GOE is another question entirely. Some argue that these bacteria—and the oxygen that they produced—could have existed for many millions of years before the GOE, but that oxygen may have been reacting too quickly with everything around it to accumulate in any significant quantity. Only once something else within the Earth system shifted around 2.4 billion years could oxygen begin to accumulate enough to leave a widespread mark in the rock record.

To support this case, some researchers point to what they think are microbial fossils in rocks as old as 3.5 billion years—roughly 1 billion years before the GOE. If these rather simple blips and blobs are cyanobacteria, then these microbes could have evolved long before the GOE and could not have been the instigator of it, as Woody argues.

To further argue their case, researchers in this camp point to some thirty locations around the world where they think they have found evidence of oxygen in rocks hundreds of millions of years older than the GOE. These geochemical traces—invisible to the naked eye but visible through laboratory analyses—support the argument that the very earliest populations of cyanobacteria could have been blooming and sending off little poofs of oxygen for a long while, but that oxygen could not accumulate because it reacted with everything around it. Only after the volume of materials available for reactions diminished could oxygen begin to build up in the atmosphere.

These so-called "whiffs of oxygen" have been a source of heated debate ever since a team of geologists based at Arizona State University first published a paper arguing for them in 2007. Woody and others in his group have worked to refute some evidence for these whiffs, arguing that they are artifacts of more recent oxygen rather than an original signal from the Archean. But those refutations have, in turn, been subject to their own refutes. "We find these arguments logically flawed and factually incomplete," write the authors of one 2023 paper in response to refutes posed by Woody's team.

Andy Knoll, Woody's PhD advisor who now holds an emeritus position at Harvard, says that some arguments against whiffs of oxygen may very well be valid.

"But if even one of those thirty things is correct," he says, "then there were cyanobacteria before the GOE."

And if that was the case, then something else must have triggered oxygen's accumulation. "My sympathies lie with the idea that cyanobacteria as a type of organism long precedes the GOE," Knoll says.

When I raised these points with Woody over our coffee chat in Pasadena, he placed his elbows on the table, took a breath, and looked me in the eye.

"We are asking rocks to tell their history, and they are imperfect archives," he told me. Studying strata is like reading the pages of a book—if you took that book and dropped it to the seafloor and then expected to be able to read everything word for word after that seafloor cracked up into a mountain range millions of years later. "It's amazing that we can say anything, and we shouldn't be too surprised that there are things that we think we can say that we are just mistaken about."

Until he sees definitive evidence otherwise, he will maintain his stance that the rise of oxygen began with the rise of cyanobacteria, presumably around the time that signs of the GOE appear in the rock record roughly 2.4 billion years ago. Sure, evidence of life does appear around 3.5 billion years ago. But those microfossils are, to his mind, indistinguishable filaments that could have come from anything with a single cell. Microbial fossils of this age have such simple morphologies that it's all but impossible to know how, exactly, they made their food or what they released as the byproduct of that metabolism. Bacteria that can't produce oxygen look, in many cases, identical to those that can in the fossil record. Woody remains wary of them not to be combative, but to underscore how disruptive false premises can be to progress in science. Barghoorn and Tyler thought they were dealing with the oldest evidence of life on the planet with the Gunflint Chert microfossils, and they thought that this oldest life was algae. But it turned out that the algae were actually bacteria and that the rocks—later dated at around 1.9 billion years old—were far younger than the actual earliest evidence of life on the planet that dates at least as far back as those 3.5-billion-year-old filaments.

They had so many details wrong, and it was only in recognizing those errors that the field was able to make any progress and grow. By latching on to his preferred GOE narrative and running with it, Woody isn't putting blinders up so much as sticking with an argument until he is proven wrong. It's only through this type of rigor and thought experimentation that anyone studying Earth's oldest rocks can hope to weed out false positives and weak links and begin to paint a clearer picture of what actually took place.

"We know, in the back of our heads, that there's one answer," Woody told me, taking a sip of coffee. "One thing happened. Something happened. So everyone can't be right."

Even so, I still couldn't help but wonder why, exactly, it was worth all this trouble trying to figure out who was right. Was it not enough to know that oxygen wasn't always here and that now, thankfully, it is?

When I voiced this concern aloud, Woody leaned back in his chair and laid it out plainly for me. With the arrival of cyanobacteria, life's tug on the planet became greater than it had ever been before. Today, as humans, we've raised the bar for global change. To understand what happens to a world when life overtakes the systems that it runs on, we can look to the past and see what has happened before us. We can try to understand how systems adjust and balance out. Who makes it and who doesn't. Who suffers and who thrives. Who gets a chance at renewal and who gets left to melt back into the mantle.

"If you get this right," Woody told me, placing his cup back down, "you know how the Earth works."

✦ ✦ ✦ ✦

BY "HOW THE EARTH WORKS," Woody meant how air, rock, water, life, and ice all interact in the web of feedback loops that geoscientists call the Earth system. Together, the five facets of this system—the atmosphere (air), lithosphere (rock), hydrosphere (water), biosphere (life), and cryosphere (ice)—orchestrate the global climate and, in turn, the underpinnings of our lives. It's by coming to understand this system that I have grown to see the physical world not as the static backdrop of our daily experience but as an ever-changing vessel that ripples and responds to innumerable changes, and has been doing so for billions of years. Over time, these subtle transformations build, erode, and rebuild the world anew. We live our lives within recycled landscapes and those recycled landscapes live within us.

I mean this literally, not figuratively. The science is the poem and the poem is the science. Everything on this planet connects with everything else, from the microscopic contents of the air we breathe to the macroscopic movements of continents and ocean currents. You can't build a mountain range without changing the atmosphere, at least a little (because freshly sculpted mountains pull carbon dioxide from the atmosphere), and you can't change the atmosphere without changing the chemistry of the ocean (because oceans absorb and release carbon dioxide), and you can't change the ocean without affecting the life within it. The loops of connection spin on and on.

Understanding the intricacies of these interconnections has become ever more critical as geoengineering gains traction as a potential antidote to our climate crisis. If we are to seriously consider tightening our grip on the planet as a way to heal it, we have a strong imperative to first understand how the planet works. The strata can tell us how the behaviors of ancient environmental engineers have rippled through the Earth system on geologic timescales, which in turn can tell us how this system might respond today if we shoot aerosols into the sky to deflect solar radiation or pump carbon into the deep sea. We can learn that if we are not mindful, our manipulations will cause more harm than good.

When Woody arrived at Harvard, the field of geobiology—the study of the interconnectedness of the Earth system—was still relatively new. But while this was a new way of thinking in modern Western science, many Eastern religions and Indigenous cultures had long recognized the cross talk between life and Earth, and how that ancient dialogue led to the evolution of the world we inhabit today.

"Evidence of our shared origin can be found all around us," writes Sherri Mitchell, an Indigenous lawyer and author from the Penobscot Nation, in her book *Sacred Instructions*. "Science has finally caught up with what we have always known, that we are all related."

Earth historians are only beginning to appreciate and understand all the intricate ways that the pieces of this system have evolved through time. It's an area of science that, itself, is evolving through time, and one that is deeply vulnerable to error. What we think is true today could be disproven tomorrow by a new rock sample or a new laboratory reading. Nothing is set in stone, because our understanding of the stones keeps changing, as do the stones themselves. This is a hazard of the job of the Earth historian, and you have to be okay with that in this line of work. Not only be okay with it, but perhaps even be motivated by it, humbled by the fact that you may never know for sure whether or not you are correct—grounded by the persistent reminder that you are small.

While the fields of physics or engineering or biomedicine provide satisfying answers to concrete questions, stratigraphy offers drafts of stories that must remain open to revisions as new techniques and new minds work to tackle them from different angles over the years. It's the humanities of the hard sciences.

The job of the stratigrapher is to always remain mindful of the ways that stories may have been rewritten over the eons, and to always be willing to admit when their interpretations of the rocks might be leading their stories astray.

Three

TO UNTANGLE ANY NARRATIVE, YOU NEED A SENSE OF chronology. What came first and what followed. You also need a sense of pacing, what happened how soon before what else. Otherwise you might see a cause and effect where there is actually a massive gap in time and no relation at all.

This, of course, proves rather difficult when you're trying to reconstruct a narrative that unfolded more than 2,000,000,000 years ago.

Today, we have a geologic timeline to help make sense of the pacing in the strata. This nested doll of subcategories decreases in size from eons, to eras, to periods, to epochs, to sub-epochs, to ages. It is a human construct that itself is still up for revision. The end of one geologic period and the beginning of another is not something that can be discovered like the Higgs boson or the double helical shape of DNA. Nor is it something that can be measured. It is a decision to be made by a group of humans.

Imagine the monumental task the world's earliest geologists took on in their efforts to find order in the strata around them, with no such timeline at hand. Where would they even begin? How could they make sense of the chapters at their fingertips when they didn't even know how many chapters there were, where the oldest one was buried, and how many were no longer visible at the surface of the earth?

The fact that Stanley Tyler knew that the chert he discovered on the shores of Lake Superior came from the time of Darwin's dilemma—and, for that matter, that Darwin even

AIR 33

knew that he had a dilemma at all—relied on groundwork laid by many other humans before them.

In the Western world, that groundwork is often attributed to James Hutton, a clever eighteenth-century Scotsman who had enough wealth and friends in high places to sway how Westerners understood their place in time and space. This doctor turned gentleman farmer turned geologist spent many years observing how sediments drifted around his farm in southeastern Scotland. With one gangly arm propped atop his hoe and his broad pale forehead crinkled into a furrow, he watched sediments blow off the peaks and fields around him and drift into streams. He could see the product of this erosion—the muddied water—but he could not see the impact of the erosion on the land. He could not perceive the peaks around him dulling nor the fields around him winnowing down. And yet he knew that they were, because these were the landmasses from which those sediments had blown.

As he roamed his farm, crinkling his long nose into awkward grimaces when deep in thought, Hutton mulled over the implications of his observations. The Bible, he knew, allowed for only about 6,000 years since the planet's creation. To his mind, this was not nearly enough time to account for all the geological processes taking place on his farm. Not a particularly religious man himself, he pushed the Bible aside and arrived at two epiphanies. First, geologic processes must take place over incomprehensibly long periods of time, far longer than the six millennia allotted to Earth within the Bible. Otherwise, he would have been able to see the landmasses around him shrinking before his eyes, and he couldn't. Second, Earth's surface materials must experience a sort of cycle of erosion and replenishment followed by more erosion and more replenishment. Otherwise, the hills and fields around him would eventually whittle down to nothing at all, which seemed quite unlikely.

He presented these theories to the Royal Society of Edinburgh in 1785, and published a paper based on those lectures in 1788.

"The result, therefore, of this physical inquiry," Hutton famously ended that 1788 manuscript, "is that we find no vestige of a beginning,—no prospect of an end."

He didn't have an easy time convincing the masses of his theories at first. As biographer Jack Repcheck writes in *The Man Who Found Time*, Hutton's deviations from the Bible earned him plenty of naysayers. His published lectures received unflattering reviews, including from one anonymous critic in a 1788 issue of the *Monthly Review* who accused Hutton of arguing for "a regular succession of Earths from all eternity! and that the succession will be repeated for ever!!"

This progenitor of the internet troll had misinterpreted the material that he mocked. Hutton had not meant that Earth had or would continue to exist forever, Repcheck notes. Hutton, instead, meant that Earth recycles its landscapes over unknowably long stretches of time. But insofar as his views ran counter to the timeline laid out in the Bible, they were largely rejected as dim-witted, impious, and even blasphemous.

Eventually, though, people came around to his ideas. Today, those final words of his manuscript echo through the lecture halls of introductory geology courses across the world. His work would later inspire Charles Lyell's *Principles of Geology*, a tome first published in 1830 that some argue was the most important book that Darwin brought with him on the HMS *Beagle* the following year. Without it, he would have had no framework around which to hang his theories of natural selection and evolution.

For all his sway, Hutton has been nicknamed the Father of Modern Geology. His contributions to the field are certainly worth noting and applauding. But was he really the very first human to understand Earth as something incomprehensibly old and cyclical in nature? Of course not.

"The ancient people perceived the world and themselves within that world as part of an ancient, continuous story of innumerable bundles of other stories," writes Laguna Pueblo poet and author Leslie Marmon Silko.

Ancient Hindu texts refer to measurements of time called kalpas,

AIR

which encompass the duration of a cosmic cycle and span some 4.32 billion years—a number astonishingly close to the actual 4.54-billion-year-old age of Earth as measured through geochemical analyses of mineral grains as recently as the mid-twentieth century. Aboriginal people of Australia—the planet's oldest continuous civilization, which has persisted for as long as 75,000 years—refer to the creation of the world as Dreamtime, a period that must date further back than the 6,000 years allotted in the Bible, considering that their civilization has existed close to 70 millennia longer than that.

Hutton earned his nickname not because he was the first person to have these thoughts of Earth's antiquity, but more likely because he was white and male—prerequisites to be taken seriously in Earth science in the eighteenth century and, by certain measures, to this day.

As the field of geology continued to expand after Hutton's epiphanies, it called for something beyond the written and spoken word to illustrate how landscapes evolve through time. Down the hall from Woody's office, a floor-to-ceiling map of the rocks of England and Wales captures another powerful way to convey long-term geologic processes and changes. With its psychedelic swirl of blues and reds and greens representing the composition and relative ages of strata across the landscape, it was the most detailed and expansive geologic map of its kind when it was printed in 1815. It allowed geologists to orient themselves in time and space with new clarity, and it continues to inform the way such maps are drawn today.

But while geologic maps may help illustrate landscape-scale changes through time, they still have limitations of their own. They can't offer anything in the way of the absolute age of the rocks. They can only tell you what came first and what followed, but not how much time spanned between events.

Absolute ages of Earth's oldest deposits became available only with the advent of radiometric dating, which came about in 1907, and the subsequent development of mass spectroscopy in the 1940s. Geochronologists today calculate the ages of the oldest rocks using zircons, tiny glassy minerals with traces of radioactive uranium. Because uranium decays into lead at a known rate, geochronologists can back-calculate the age of a zircon based on its ratio of uranium to lead. It is this relatively recent advent of radiometric dating that has proven key to establishing the pacing of the Great Oxygenation Event, and so many other events in Earth history.

It's how we know that sharks are older than the North Star, eyes are older than leaves, horseshoe crabs appeared hundreds of millions of years before flowers, and flowers showed up dozens of millions of years before grasses.

It's how we know that, for more than half of Earth's existence, there was no oxygen in the atmosphere, and for billions of years beyond that, there were no lungs to breathe it in.

✦ ✦ ✦ ✦

NOT LONG AFTER BARGHOORN and Tyler discovered their groundbreaking fossils in the Gunflint Chert, Canadian geoscientist Stu Roscoe made another important discovery that helped track oxygen's initial rise in the atmosphere. This work remains foundational to GOE research today.

A skilled field geologist, Roscoe was also a seasoned bush pilot notorious for crashing or nearly crashing his plane. Through his position with the Mineral Resources Division of the Geological Survey of Canada, Roscoe spent his days flying across northern North America inspecting potential mineral prospects to advise the Canadian government on their viability. Along the way, he'd routinely leave his gas tank dangerously low and take dips and turns that churned the stomachs of his passengers.

Once, while en route to fill up on fuel in the Northwest Territories, he and a colleague got so distracted taking pictures of a scenic waterfall that they didn't notice their fuel gauges swing to empty. Not sure if they'd make it to their destination alive, his colleague

AIR

reclined his seat as far as it would go and took a picture of the gauges. "I thought that if we survived, I would make it into a Christmas card to regale the Division," the colleague later wrote. When they geared up to land, the engine coughed and then stopped. But just before the plane hit water, enough fuel sloshed forward to catch the engine and carry them to their destination.

This kind of blind luck seemed to follow Roscoe around the tundra and taiga where he worked. When he wasn't careening around making others nervous on planes, he was mapping rocks within the Canadian Shield, an expansive region centered on Hudson Bay that contains some of the oldest strata in the world. Among those deposits, Roscoe found one collection of sediments that struck him as particularly unusual. Within remnants of an ancient riverbed, he found sand grains of uranium. Grains that had clanked and clattered against one another as they toppled downstream, submerged in water for long periods of time.

Roscoe probably blinked hard when he found those deposits. He would have known that such grains do not exist in today's world because uranium dissolves in water—or, at least, it does in water with oxygen dissolved within it. For those minerals to have tumbled downriver as grains of sand and persisted long enough to get locked in the sediment, something had to have been very different about the contents of the atmosphere at that time. "It leads . . . to a consideration," he wrote in a 1969 report, "of the possibility that the deposits were formed at a time when the earth's atmosphere contained no free oxygen and that similar important deposits may not be found in younger rocks."

"It's easily the most important observation for oxygen," Woody says of grains like these, which pop up elsewhere around the world in rocks older than the GOE, and then disappear after it. He and his colleagues continue to track these proxies as they carry on collecting field observations to firm up the timing and series of events that spurred the GOE. "We know what these minerals are, we know how they behave. Anyone can repeat my work if I give you a magnifying glass. We understand the mechanics of it more or less perfectly, that's what makes it so powerful."

Back in his office at Caltech, I came across another similarly powerful proxy of oxygen's early absence in the atmosphere that also continues to inform research on the GOE.

I shuffled around a drone in a cardboard box and a whiteboard full of colorful scribblings to get to Woody's desk, where his wife and young blond son grinned out from a small silver frame. Just beyond the photograph, I found a petri dish filled with polished stone discs the size of poker chips. A few were black with subtle swirls, another was gray with red streaks. But the one that caught my eye had dozens of glimmering, gold-hued grains the size of poppyseeds, trapped within a dark brown backdrop.

I lifted it between my thumb and forefinger and drew it close to my face. I had never seen a rock like this before. I tried to read it, figure out how it had formed.

The metallic grains had a sort of fluid flow to them, as if water had rushed over and arranged them into a mess of ripples. The bed of a river appeared in my mind's eye. The shapes of the grains pointed to rushing water as well—some had jagged edges but others were smoothed into spheres, the result of bumping and smacking and hitting other hard things over extended periods of time. I imagined the grains jumping along the bottom of a channel, pounding and flicking flecks that progressively grew smaller and smoother on their way downstream.

I recognized those grains as pyrite, fool's gold. My grandfather gifted me a chunk as a child and I had taunted my brothers with it, telling them I had struck gold in our backyard but knowing that its luster wasn't quite as warm as real gold.

Woody confirmed my hunch.

Like uranium, pyrite rapidly disintegrates to nothing when exposed to water in today's world. It can rest in museum display cases and gift shops, but even then will begin to rust if exposed to too much moisture.

"You look at that," Woody told me as I shifted the smooth disc back and forth between my fingers, "and you're like, 'That is not a standard rock type. I bet the Earth was different then.'"

The rock had come from South Africa, one of the few places in the world where deposits from before the GOE still remain at Earth's surface. The disc was a portal to a time when no eyes would have seen those warm buttery grains, how they glimmered in the trickles that washed over them.

Though proxies like these can help track the duration of oxygen's absence and the timing of its arrival, they alone can't reveal anything about why oxygen arrived when it did. Reconstructing this narrative also requires a solid sense of the characters at play. Who they are, what roles they fill, and—most important—at what point they arrive onstage.

✦ ✦ ✦ ✦

USHA LINGAPPA JOINED WOODY'S team as a graduate student in 2015 to help clarify the players of the GOE.

"There is a kind of scientifically credible creation story vibe in doing this work, and it appeals on a very deep level because of that," she told me over a video chat from her office at UC Berkeley where she was working as a postdoctoral researcher.

Her research questions are as much scientific as they are existential. How did we all get here? What series of events made our existence possible? While her answers may arise through the scientific method, they retain an element of mythicism in the way that a creation story might, because so much remains unknown and, to a certain extent, unknowable.

Usha started out as an art student in college and then pivoted to a self-designed major in astrobiology with a focus on the origins of life that, in turn, led her to study the origins of oxygen-producing photosynthesis with Woody. She is one of a number of Woody's students who actively practice art alongside their science. "You make connections that you wouldn't otherwise make when you look at the world through a more artistic lens," Usha told me. Indeed, studying Earth's earliest days calls for a type of three-dimensional, creative thinking more often associated with the visual arts than with the sciences.

Going into her PhD project, Usha understood that the earliest oxygen-producing bacteria would have needed to solve two problems at once. They would have needed to harness the enormous amount of energy required to break water molecules apart, and they would have needed to cope with the dangerously high reactivity of the oxygen they produced as a result—reactivity that could mutilate cellular machinery and create poisons like hydrogen peroxide.

"Oxygen is toxic," Usha reminded me with eyebrows raised. Evolving the capacity to produce oxygen would have been a rather dangerous enterprise. "So how did they evolve it, and how did they survive evolving it?"

She spent her six years at Caltech digging into these two questions that, as she discovered, seem to share the same answer.

Early on in her project, she traveled with Woody to the University of Johannesburg in South Africa to look through rock samples collected from the Kaapvaal Craton—the ancient heart of an old continental plate with mineral ages ranging from about 3.6 billion to 2.5 billion years old. As one of the very few locations in the world with rocks dating so far back, this craton has been integral to investigating the GOE.

The rocks of interest had been collected as pieces of drill core, long cylinders of stone pulled from the earth with a giant drill bit, sort of like an apple corer. Drill cores provide long, uninterrupted views of strata within crust not otherwise visible at Earth's surface. They are invaluable to weaving together stories from Earth history, particularly from the Archean—4 billion to 2.5 billion years ago— since only a small fraction of rocks that old remain at the surface of the planet today. One estimate suggests that Archean deposits make up less than 3 percent of Earth's modern surface. The pool of material to pull from expands quite a bit when you can access strata deeper in the crust.

The cores from the Kaapvaal Craton measured longer than a football field in length, but they had been cut into more manageable slices a few feet long, and then filed away in dozens of metal boxes in a shed on campus. Usha and Woody gathered those boxes and laid them out on a patch of sunny pavement to take stock of what they were working with. The drab gray cylinders that lay before them would have looked rather unremarkable to a passerby, but their ash-white and sooty swirls mixed with bits of rusty red layers held far more intrigue than met the naked eye. Usha and Woody could squint past their sepia tones and begin to color in the hues of the GOE.

They had not seen this particular core before, so they didn't know what exactly they would find. They did know that it had been collected not far from a defunct manganese mine. A bit beyond the mine lay the Kalahari Manganese Field, the site of some 77 percent of the world's land-based manganese deposits. Concentrations of this metal are so unusually high in this area that calves and lambs grazing in the region suffer from a rare manganese-poisoning disease that has not been found anywhere else in the world. That Archean manganese layer, it seems, represents a singular moment in geologic time. If that layer extended out into the core, Usha and Woody reasoned, then it could prove key to understanding the origins of oxygen.

This was not an arbitrary connection based solely on the proximity of the manganese to the core. This metal plays an integral role in oxygen production today. It clusters up in chloroplasts—the cellular location where photosynthesis takes place—and serves as a sort of solar panel that stockpiles the staggering amounts of energy required to break apart the stubborn hydrogen and oxygen bonds within water. Manganese also has the power to shield living things from the toxins produced by oxygen. The bacterium *Deinococcus radiodurans* can survive radiation doses 1,000 times greater than we can, and they do so by stuffing themselves full of manganese. We use this metal in our own bodies as well, to protect against cancer-causing free radicals. Across the tree of life, you can correlate radiation resistance to the manganese content of cells, Usha told me.

She and Woody thought that manganese may have offered a panacea for the dual challenges posed by generating oxygen. It could have provided the energy boost needed for early microbes to create oxygen, and the protective shield needed to survive creating it. What's more, this metal would have been abundantly available within Archean oceans, with concentrations of dissolved manganese possibly 1,000-fold greater back then as compared to now. All of that dissolved manganese would have gotten into those ancient oceans by eroding off continents, as it does today. But in the modern world, this metal can't remain dissolved for long because it reacts with oxygen and falls to the seafloor as a solid mineral grain. In the Archean, however, it would have had no such obstacle and would have accumulated in waterways unabated.

"This would have been a pretty appealing ecological opportunity," Usha told me.

Give a honeybee a flower and she will make honey. Give a cyanobacterium enough dissolved manganese, and perhaps it will make oxygen.

It's one thing to wonder if ancient microbes took advantage of this ecological opportunity, but it's another thing to figure out how to find evidence of this in the rock record.

To their great delight, Usha and Woody found a gleaming breadcrumb of a clue on that patch of pavement in Johannesburg. They easily identified the manganese layer as a strip of inky-black minerals far darker than the rest of the gray core. But it wasn't until they wet the rock down to see the patterns more clearly that the significance of that dark layer came into focus. There within the manganese they found the familiar crinkled layers of microbial mats piled into dome-shaped mounds. They were stromatolites, those same structures that Stanley Tyler had found in the Gunflint Chert.

To their knowledge, nothing like this had ever been described before. Stromatolites are pretty common in the rock record, but stromatolites encased in this amount of manganese are not. There had to be some reason, beyond mere coincidence, that these microbes lived so intimately enmeshed with this metal.

"It definitely wasn't lost on me that they were once-in-a-lifetime samples," Usha told me.

Throughout the core, she and Woody identified several telltale proxies that confirmed that they were straddling the GOE. They found riverbed grains of pyrite and uranium that disappeared about a third of the way up the core, and the gradual appearance of rusty red layers toward the top. It's in this transition into the red rocks where the GOE falls.

"We can almost put our finger on the horizon where it exists," Woody says.

The proximity of the manganese-loving cyanobacteria to the rusty red strata within the core suggests that they were growing right around the time that the first oxygen began accumulating in the atmosphere. These were, to Usha and Woody's minds, either remnants of some of the earliest oxygen-producing cyanobacteria or progenitors of oxygen-producing cyanobacteria that somehow laid the evolutionary groundwork that eventually made oxygen production possible.

After three days of observing, measuring, and sketching the core, followed by a week or so out in the field to gain a broader context for the rocks, Usha nervously loaded the samples into her luggage and flew home with them.

Back in the lab at Caltech, she scanned the stromatolites using X-ray fluorescence, a technique that identifies chemical constituents within rocks. The results astounded her. The space in and around the ripply microbial mats glowed with far more manganese than she had expected. There wasn't just a trace amount—the microbes were so encased in the metal, they looked like they had actively hoarded it.

"When I saw them," Usha said, "my heart stopped."

She conducted follow-up studies in her lab with modern cyano-bacteria to see if manganese hoarding was really something these microbes were capable of doing. She found that it was. Maybe, Usha reasoned, the ancestors of modern cyanobacteria did this as well. Maybe they built themselves a solar panel of manganese molecules that allowed them to capture enough energy from the sun to break apart water, and maybe that manganese also allowed them to survive their own toxic haze of oxygen that they produced as a result.

She shared her findings with colleagues in 2021, in a video presentation complete with her own watercolor illustrations depicting the intertwining relationship of electrons and oceans and continents. "These are the communities that used manganese to unlock oxygenic photosynthesis and, in so doing, change the fate of our planet forever," she told them.

Whether those South African microbes were the first photosynthesizers to break apart water molecules, or whether they instead provided an evolutionary stepping stone to do so, still remains unclear. But it appears likely to Usha that this metal played a central role in the early rise of oxygen-producing photosynthesis. Without all of that manganese eroding off Archean landmasses and accumulating in those oceans, we might not be here today. Earth's crust created us.

"The co-history of cyanobacteria, manganese, and oxygen is an incredibly important and firm thing, especially after Usha's work," Woody says. Once oxygen started pooling up in the atmosphere, the excess manganese in the oceans would have pretty quickly rained out as solid grains on the seafloor. Ever since then, dissolved manganese would have never had the chance to accumulate in the same concentrations, which is why we don't see deposits like this before or after. "You have a little sweet spot in time," Woody says.

Agonizingly, though, they can't pinpoint the exact date of that sweet spot because they haven't found datable material within the core that they studied from the Kaapvaal Craton. They need zircons, those tiny glassy minerals with traces of radioactive uranium

AIR

45

that geologists use to calculate the ages of Earth's oldest rocks. Zircons from material collected nearby the core offered up dates of roughly 2.5 billion years, so they knew that they were working roughly within that vicinity. But they can't say for sure how old the manganese-laden stromatolites are.

That's a bit unfortunate, since dates are key to tightening up the timing and tempo of the GOE and for making arguments either for or against whiffs of oxygen before it. For whiffs to exist, oxygen-producing photosynthesis would have needed to exist. But if this manganese layer really does represent a singular moment in time before which oxygen-producing microbes could not have existed, then that would certainly weaken the argument for earlier whiffs.

Whatever the case may be, these manganese-loving microbes did not evolve in a vacuum. They grew floating in turbulent oceans that lapped against barren continents that hung over a molten mantle that spun around Earth's maturing core. To fill out the rest of oxygen's narrative—and gain a stronger understanding of the Earth system—we must consider what else was happening in the world that could have spurred the growth of these characters at this time.

This is where the plot thickens: where the oceans and the continents and the atmosphere look like they may have conspired to help land cyanobacteria center stage.

✦ ✦ ✦ ✦

IN THE EARLIEST OF Archean days, Earth's hot young interior would have kept the crust from cooling and stabilizing in the way it has today. The cookie dough of the continents had not yet fully set, so landmasses melted and dripped their bottoms back into the molten mantle as unstable crustal pieces. The source of this heat remains up for some debate, but the consensus points to a combination of radioactive decay within the lower mantle and primordial warmth generated by the creation of the planet—both of which have naturally dissipated as Earth has aged. We know this thanks, in part, to a rare type of dark gray or greenish volcanic rock called komatiite that all but disappears after the Archean. Geochemical analyses of these ancient magmas indicate that they flowed at temperatures far hotter than magmas of the modern world.

Earth's early heat also sent moisture in the mantle sizzling up into the oceans, likely flooding marine basins with far more water than they hold today.

In the absence of stable continents and the presence of vaster oceans, most things on this young planet—both geologic and biologic—existed under the sea. Geologists make this assumption based on theory and on actual evidence in the rock record that shows a preponderance of underwater volcanoes and a near absence of ones that erupted above water during this time.

Then, between about 2.7 billion and 2.5 billion years ago, something changed. The planet began building more continental crust than it ever had before. For the first time, Earth had large expanses of rock exposed to wind, rain, ice, and snow. It's a change so pronounced in the rock record that it has been invoked as the rationale for defining the end of the Archean at around 2.5 billion years ago. The shift into the next eon, the Proterozoic, marks the beginning of a planet with this cooler, more stable crust.

Maybe, some researchers argue, all that oxygen needed to accumulate was for the young Earth to finally mature and chill out. Just as the human body morphs from the inside out as it matures through adolescence, so too has the planet morphed from the inside out as it has grown older. But how would these changes influence the ways that gases accumulated in the atmosphere?

For any gas to accumulate, its sources (the things that billow it up into the atmosphere) must overwhelm its sinks (the things that suck it down into the earth). If the sources can't keep up with the sinks, then the gas will swiftly seep away like liquid through an unplugged drain.

When oxygen first arrived, its sources would have been limited

AIR

47

to the one group of microscopic bacteria that created it. Its sinks, on the other hand, would have been far more expansive. From the seafloor to the stratosphere, ions and gases and minerals and other molecules would have accepted oxygen's plea to connect and then incorporated it into their solid and liquid and gaseous forms. Only once it filled these sinks could oxygen finally begin to overflow into the atmosphere. For this to have happened, either the populations of the microbes that created this gas would have needed to proliferate and make more oxygen, or oxygen's sinks would have needed to shrink—or, perhaps, some combination of the two would have needed to take place in tandem.

Woody and others in his camp see oxygen's sources as the lead drivers of the GOE, and any shrinking of its sinks as secondary to the plotline. Those who argue for pre-GOE whiffs, on the other hand, think the opposite—that oxygen's fate rested in a fundamental change to its sinks as continents solidified at the end of the Archean.

This shift in Earth's crust could have shrunk oxygen's sinks in a number of different ways. Scan the academic literature on this and you'll find more than twenty different rationales. Some argue that gases that spew from submarine volcanoes react more readily with oxygen than gases released by volcanoes on land. If most of the planet's landmasses sat underwater through the end of the Archean, then most of the volcanoes would have erupted underwater, and the total volume of gases reactive with oxygen would have been rather high. Once the planet's primordial heat dissipated and the continents began to stabilize, more volcanoes would have erupted above water and oxygen's sinks would have shrunk.

Lee Kump, a geochemist at Pennsylvania State University and coauthor of the textbook *The Earth System*, made this case in a 2007 paper and stands by it today. He points to it as one potential explanation for whiffs of oxygen prior to the GOE. Cyanobacteria could have existed long before the GOE, but they only left their global mark once the planet matured and hardened up.

This is his preferred theory, but he has not entirely ruled out the simpler explanation that Woody advocates for. "I'm perfectly happy with either of those explanations ultimately winning," he told me over a phone call. "I don't think we have wasted or derailed science either way; it's a really important thing to try to understand."

Aleisha Johnson, a geochemist at the University of Arizona who works with the team that originally coined the term "whiffs of oxygen," agrees that there is still room for debate, but she sits rather firmly in the whiff camp. "I find that line of argument most convincing," she told me.

These arguments have opened up new opportunity for collaboration across different fields of geoscience. The debates around oxygen's origins are helping to fine-tune our understanding of the underpinnings of the Earth system so that we may tackle larger looming arguments, such as how to address the climate crisis today. "I couldn't say I am solving climate change," Johnson told me, "but I do think it will broadly help if we have a better understanding of just how the Earth works as a whole with these large perturbations."

However you slice it, the rock record tells us that something fundamentally changed inside and outside the earth some 2.5 billion years ago, give or take a few hundred million years. Only by digging deeper into the well of strata available from this time may we begin to spin up a clearer understanding of how Earth's insides shaped its outsides and landed us in our current moment today.

Four

IF YOU'RE LOOKING TO IMMERSE YOURSELF IN A TIME leading up to the GOE, the Soudan Iron Mine in northeastern Minnesota allows you to do just that. What once was a leading producer of US iron ore now offers educational public tours through chambers that descend nearly half a mile inside the earth, within rocks aged around 2.7 billion years old.

Back when the mine was in operation from the mid-1880s through the early 1960s, its ore was smelted into steel that got hammered into railroads, bridges, airplanes, and countless other bits of infrastructure that helped shape the modernizing country. At the mine's peak, some say, there wasn't steel made in the United States that didn't have at least some trace of Soudan iron in it.

But by the early 1960s, when more lucrative ore bodies were discovered elsewhere, the U.S. Steel Corporation donated the mine and the surrounding 1,200 acres to the state. Soon after, the facility opened back up for public underground tours as the Lake Vermilion-Soudan Underground Mine State Park. The park now welcomes tens of thousands of visitors every year who travel into the earth to learn more about the mining history of the region and to commune with the Archean seafloor that comprises the ore.

Iron deposits like these generally exist only in rocks that formed before Earth became fully oxygenated, with a few discrete instances when they reappear later on when oxygen levels dip. Like dissolved manganese, dissolved iron falls to the seafloor as a solid mineral grain in the presence of dissolved oxygen. Before the GOE, the world's oxygen-free oceans swirled with enormous volumes of dissolved iron that emanated as gas from vents in the seafloor and eroded off bedrock. Some intermittent chemical reactions caused some of that iron to fall to the seafloor as a solid, but large volumes remained dissolved. Once enough oxygen accumulated in the atmosphere and permeated the oceans, however, dissolved iron would never accumulate in such volumes again. Today, a single liter of ocean water may contain only a billionth of a gram of dissolved iron; deposits of iron ore indicate that concentrations were magnitudes larger than that prior to global oxygenation.

Had it not been for those pre-oxygen days, these iron-rich rocks would not exist because dissolved iron would never have accumulated in such high concentrations. The production of steel would not be what it is today, nor would the global economy. We need there to be oxygen in the atmosphere, but we also need there not to have been oxygen in the atmosphere in order to exist in the modern world the way we do, with steel cars and kitchen appliances and medical devices and airplanes and so on.

I arranged to visit some of those pre-GOE iron deposits in Minnesota to get a better handle on what that Archean world looked like, how it resurfaces in our lives today, and what it might tell us about the onset of the GOE.

✦ ✦ ✦ ✦

MY TOUR GUIDE THROUGH Minnesota's iron deposits was Amy Radakovich Block, a geologist with the Minnesota Geological Survey who specializes in mapping the state's oldest geologic units. As one of the people most familiar with some of the country's oldest rocks, she was well positioned to give me a sense of what the world looked like in the few hundred million years leading up to the GOE, and what questions still remain about this time.

I met Amy in the parking lot of her brick office building in St. Paul on a clear June morning. She wore water-resistant pants tucked

AIR   51

into wool socks and a light brown ponytail neatly arranged under a
ball cap, ready to spend the day in the field.

She had jam-packed our schedule. On our way to the Soudan
Iron Mine we would stop at a few outcrops to look at evidence of a
maturing crust leading up to the GOE. Then, once inside the mine,
we'd acquaint ourselves with iron deposits laid down in Archean
oceans, and witness some of the challenges these rocks pose in set-
tling the debates around the whiffs of oxygen. The next day, we
would meet up with Latisha Brengman, a colleague from the University of Minnesota Duluth who studies the geochemistry of iron
deposits like those in the mine. She is conducting cutting-edge geochemical analyses that may help settle the whiff debates.

Unlike Woody and his colleagues whose careers revolve around
the GOE, Latisha and Amy are more agnostic in their study of these
rocks. Latisha is most interested in the underlying geochemistry
expressed within the strata rather than the GOE more broadly, and
Amy's research centers more on identifying the rocks than ascribing
any particular story to them. Traveling with the two of them would
offer a neutral perspective on the strata and what they may or may
not say about the rise of oxygen. We'd get a sense of how hard it can
be to tease out legible details from deposits that have been torn and
mangled over billions of years, and how to find the most reliable
traces of narrative still left intact.

To get to the Soudan Iron Mine from St. Paul, you travel about
250 miles north along pine-lined highways, passing towns with
names like Sandstone, Cotton, and Spruce that gradually transition
to ones like Iron Junction and Mountain Iron the farther north you
go. Amy grew up amid place names like these in nearby northern
Wisconsin and Upper Michigan, where great-grandparents on both
sides of her family worked in iron mines. Their lives revolved around
the metal and the metal revolved around them.

Before white settlers arrived here and named these towns after
the resources that they sought, members of the Bois Forte Band of
Ojibwe lived throughout what is now called Minnesota, including

along the shores of the lake where the Soudan Iron Mine now sits. They named that body of water Onamanii-zaaga'igan. When French fur traders arrived in the 1600s, they translated that name to Lake Vermilion, as it appears on maps today.

White settlers continued to trickle through the region for a couple of centuries, but it wasn't until the end of the Civil War, when rumors of mineral wealth began to spread, that the settlers appeared in droves. During the summer of 1865, just months after the Civil War ended, Minnesota's governor hired a team of two geologist brothers to prospect the area around Lake Vermilion to see if those rumors held true. The brothers did, indeed, find minerals of interest, including sizable iron deposits around the lake. But far more tantalizing to them at that time were the glints of silver and gold that they reported within quartz veins along the lake's western shore. Word of these precious metals rapidly spread south to St. Paul, sparking fantasies of a gold rush comparable to California's.

"The eye takes in a score or more graceful promontories ... running their moss and pine covered points down into the Lake, as if to cool their metallic palates already burning to be unbosomed by the restless hands of enterprising miners," wrote Ossian Euclid Dodge, a journalist for the *St. Paul Pioneer* who published a series of stories on the Minnesotan gold fever under the pen name Oro Fino, before becoming a stakeholder in the industry himself.

Over the course of that summer and fall, excitement around the prospects of a rush continued to grow. Henry Eames, one of the brothers hired by the governor, promised that the land would become a second California, and investors took note. As anticipation grew, so too did the dismay of the members of the Bois Forte Band who lived along those shores and insisted that they had never ceded access to their lands. The US government disagreed and to argue its case pointed to a treaty that it had drafted the decade prior. But tribal leaders pushed back, reminding government agents that the treaty had never been finalized.

In February 1866, a group of six tribal leaders traveled by steamer

AIR

and train over hundreds of miles to try to settle this dispute at the US capital. As has been the fate of so many disputes between colonizer and colonized, the US government did not grant those tribal leaders what was rightly theirs. Members of the Bois Forte Band lost access to the land around Lake Vermilion and were, instead, promised funds and acreage elsewhere, dozens of miles away from those shores.

With the official go-ahead from the US government, gold rushers hustled north to begin their digging, and investors as far away as Chicago and New York helped foot the bills. All seemed to be going as planned as hundreds of gold rushers set up camp and hacked away at the ground, watching the sunset redden over those waters and dreaming of what might turn up beneath their feet.

But by the end of that first season, vanishingly little gold materialized. The prospects had either been grossly overestimated or intentionally blown out of proportion. Whatever the case may have been, the gold fever broke as quickly as it had caught on. In its wake, the gold rush left behind a newly built twelve-foot-wide road that would, forever thereafter, open up the region to colonization and resource extraction. Over time, Minnesota would become the country's leading producer of iron that, in turn, would transform the US economy through the production of steel.

The Bois Forte Band reestablished a reservation around Lake Vermilion in 1881 as the result of an executive order, but on only 1,000 acres of the tens of thousands that were rightfully theirs. Members of the Bois Forte Band continue to live on these lands today, where they have built the Bois Forte Heritage Center and Cultural Museum to share their story with visitors.

"Most resources that talk about us reference us to things of the past, black and white photos, people only in regalia, or stop talking about us once the mining industry started," Jaylen Strong, the band's tribal historic preservation officer, told me in an email. "It's almost never in a modern context, so it's important to have that part of the story told."

Since opening in 2002, the Bois Forte Heritage Center and Cultural Museum has received recognition for their work from groups including the National Trust for Historic Preservation and the National Association for Interpretation. Their story of forced removal in the name of mineral exploration and economic growth is not unique to this corner of the country or the iron industry. The extraction of mineral resources has deep colonial roots across the US and beyond, as Lauret Savoy documents in her book *Trace: Memory, History, Race and the American Landscape.* "Any attempt to disentangle the search for copper or iron (or nickel or gold) from those first scientific studies, from treated dispossessings, or from early American ethnography is a fool's errand," writes Savoy. Many geologic maps in use today were originally drawn with the explicit purpose of locating mineral wealth, at the expense of the Indigenous people who lived atop it.

"There's a lot to grapple with," Amy told me as we drove north, buckled into an SUV made of metals with their own history of injustices, spewing fossil fuels bound to spill those injustices far into the future. "We have committed harms against the tribal communities. We actively caused the displacement of Native peoples."

In 2019, the Minnesota Geological Survey took steps toward drafting a policy that required its staff to consult with tribal members before conducting geologic mapping on tribal land. The following year, jolted by the murder of George Floyd just miles from its St. Paul offices, its staff doubled down on these efforts. In 2022, it became one of the first state geologic surveys to establish a policy that requires tribal consent before conducting work on tribal land both on the ground and through satellite analyses. This marks one obvious and necessary step forward, Amy told me. But she knows that they still have a lot more work cut out for them to heal their relationship with the tribes.

Some of that work will come by helping to make the geosciences more accessible to Native students, so that they have more of a voice in the field moving forward. Of the 52,250 people who earned research doctorates in the US in 2021, only 105 were Native—and a smaller fraction still were in geosciences.

"It's a hard space to occupy and maintain the duality of your identity as an Indigenous person and as a scientist," says Wendy F. K'ah Skaahluwaa Todd, a geoscientist and oceanographer at the University of Minnesota Duluth who is Alaska Native Haida.

In 2021, Todd founded the Indigenous Geoscience Community (IGC) to work to provide the mentorship and support needed to increase representation and leadership among Indigenous scholars in the geosciences and to provide a platform to meld Traditional Knowledge with Western knowledge. Beyond the IGC, Todd has worked to bridge the divide between Indigenous and Western scientists in a number of other ways, including by collaborating with a muralist on a project in Seattle that depicts three Haida women who represent air, water, and earth, with a caption that reads, "*Áajii 'wáadluwaan uu gud ahl Ḵíiwaagang*," or, "Everything is connected."

Todd is drawn to her work as a geochemist precisely for its resonance with this common Haida phrase. "That's why I like geoscience," she told me. "It's all connected."

Amy hopes that agencies like the Minnesota Geological Survey will be able to contribute to the work of groups such as the IGC by conducting outreach among tribal youth to help them view the geosciences as a field where they belong. "That's the long game," she told me as we rattled north on I-35.

✦ ✦ ✦ ✦

AFTER THREE HOURS ON the road, we pulled off the highway to hike to our first outcrop. This was the oldest of the rock we would be touring that day, though all the material we would be looking at hovered around 2.7 billion years old.

The low whines and wagging bodies of mosquitoes reminded Amy why she usually starts her fieldwork toward the end of the summer rather than in June. She sprayed her red canvas work vest in repellent, I threw a net over my face, and we walked together up an overgrown dirt road to a small clearing of poplar and pine trees. At the top of the grassy hill, the path gave way to pale gray stone with a subtle network of streaks and cracks.

"This is basalt," Amy told me, bending down to kneel on the rock that lay flat as pavement beneath our feet. This volcanic rock makes up the majority of the seafloor.

With every step we took across the outcrop, we slipped forward in time, crossing three distinct lava flows that each measured about twenty to thirty feet thick and probably erupted over the course of a few weeks or months. Without Amy's help, I never would have been able to distinguish the different flows—at first glance, the rock looked like a large and featureless stretch of pavement. But upon closer inspection, we found a handful of long, linear beige streaks that were lighter in color than the rest of the rock. Each represented the top of a flow, where molten magma had once hit cold Archean seawater and turned to glass in a flash. Over the 2.7 billion years since then, that glass had transformed into a gunky, beige rind.

Clustered toward the top of the flows were subtle bloops and blops roughly the size of basketballs, outlined with more of the same beige glass. These, Amy told me, were magmatic "burps" called pillows that had inflated from the leading edge of a volcanic flow and crackled as they hit that cold seawater. Smaller versions are, delightfully, called buns, and larger ones are called mattresses.

These flows had once lain flat on the bottom of the ocean, but had since been pushed up by tectonic forces to sit nearly vertically—though we were looking at them as a flat, horizontal surface because billions of years of erosion had filed them down to the level of the surrounding grassy hill. "The way you're standing right there," Amy said, pointing to my feet, "you're technically lying on the seafloor." I imagined the rock rising up out of the ground behind me, pressing flat against my back.

Those pillows grounded us firmly in the Archean. Their signifi-

AIR   57

cance to the story of the evolving crust of that time—and the onset of the GOE—would only become apparent at our next stop.

We drove a short distance to another trailhead that led us through a marshy forest with even thicker clouds of mosquitoes. At first, the pebbles and rocks scattered beneath our feet had the same grayish blue tones of the basalts we had seen earlier. But, about a mile in, pops of red and black underfoot hinted at what lay ahead. We continued past freshly unfurled ferns and crimson columbine flowers, over a wooden footbridge and alongside a beaver pond. We watched the white tail of a deer bounce deeper into the forest.

Finally, another mile in, Amy paused to look at her map and beep around on her handheld GPS device. "This is it," she told me as she ducked beneath a thick clump of branches and began bushwhacking off trail. I crunched in behind her, clambering over downed branches and trying not to trip over slippery leaves. About fifty feet from the trail, Amy dropped her hammer down on the stone beneath us and confirmed that we had landed in the right spot.

She reached out her hand and placed the fresh chunk of rock in my palm. It sat dense and heavy as I turned it between my fingertips. I held it up to the dappled sunlight and found distinct red and black stripes that I was familiar with from textbooks and university collections, but had never seen out in the wild. The pattern was almost zebra-like, or like the swirls of a fingerprint. I looked beneath my feet and found more alternating layers of red and black stripes. Banded iron formation, or BIF.

Amy handed over a thin yellow rod with a magnetic end that had been dangling around her neck, a simple tool that geologists use to test for magnetism and the presence of certain iron-rich minerals. As I hung the rod over my open palm, its magnetic end flicked dutifully toward the stone.

Unlike basalt, which has been oozing out of the seafloor since close to Earth's beginnings and continues to do so today, banded iron formations are an extinct rock type. New BIFs haven't formed since the oceans became fully oxygenated, save for a few isolated instances when oxygen levels dipped for relatively short periods of time. Earth will never regain the crust of its youth. And yet here, in this shady corner of Minnesota, we can touch fragments of those earlier days.

When I first learned of these striking black and red formations in the mid-2000s, the prevailing view was that they were a leading proxy for the GOE—that they recorded *the* moment of Earth's earliest oxygenation. The theory went that, when oxygen arrived, it would have quickly transformed iron into solid mineral grains that rained down to the seafloor as the red bands we see in BIFs. Once the oceans fully oxygenated, they wouldn't support nearly the same volumes of dissolved iron they once had (as was the case with dissolved manganese) and thus wouldn't have had the capacity to form BIFs.

The story was a neat one. But some parts of it turned out to be incorrect. Red herrings abound.

It was true that ancient oceans must have pulsed with far greater volumes of dissolved iron than oceans do today. This had to have been the case to explain the sheer volume of iron ore found in rocks older than a certain age and the absence of such volumes thereafter. But what wasn't true, geochemists had since discovered, was that dissolved iron needed oxygen to turn into solid mineral grains. The geochemistry of BIFs remains an open area of research, but it's clear that they don't need oxygen to form. This means that BIFs could have been carpeting the seafloor long before the arrival of oxygen. And as for that red color associated with rust and oxygen? It could have come along and stained the rock at any point after its original formation.

The role of BIFs in the narrative of oxygen's rise has since become more obscure. Rather than indicating the exact moment of oxygenation, they instead represent a time when there was far more iron dissolved in the world's oceans due to an absence or near absence of oxygen.

AIR 59

Amy's map indicated that an entirely different rock type extended in the opposite direction down the hill, close to where we stood. I wasn't convinced we'd have any luck finding the contact where the two rock types met, given all the leaf litter beneath our feet. But not long after we began our search, something beneath the leaves caught Amy's eye. She dropped her hammer and grabbed a fresh piece of rock to get a better look.

In her hand lay a light gray stone with small, tightly packed crystal grains that glistened like glass. They were bits of quartz, composed of silica. Whereas the iron formations had formed on a deep seafloor, these glassy grains pointed to a volcanic source that had erupted closer to land, in the presence of landmasses that had cooled and evolved enough to form continents.

In that very spot, along that contact, we found evidence of a seascape that had transformed into a landscape—a transformation made possible by the movements of tectonic plates. The contact between those two rock types, buried beneath all that detritus, offered evidence of a maturing crust with cool and stable continents. Quartz-rich magmas like those of the volcanic rock we stood on form from recycled oceanic crust, and such recycling requires tectonic activity.

"It's a big deal to have something this definitive from the Archean," Amy told me, referring to the exposed contact between the two types of rock.

It's not the only evidence, or even the best evidence, of Earth's maturing tectonic system, but it's an example of how traces of these pivotal geologic transitions lie hidden beneath our feet in the most unassuming locations.

Today, the planet's crust consists of some twenty plates that float above the mantle. Those plates grind past and sometimes slam into or drift apart from one another, creating mountains and valleys and everything in between. That contact in those Minnesotan woods provided a marker of the early development of the topography that we build our lives on today, the beginning of the possibility of every trail lined with beaver-dammed lakes and every white tail receding into the shadows.

✦ ✦ ✦ ✦

OUR LAST STOP OF the day brought us to the Lake Vermilion-Soudan Underground Mine State Park headquarters, where we would embark on our tour inside the earth. A middle-aged mustached man named Jim Essig greeted us wearing steel-toed boots and two layers of plaid flannel shirts. "I don't usually dress much like one," he told us as he welcomed us in, "but I am the park manager."

He gathered us hard hats and headlamps and led the way outside to the mine shaft, with a swagger leftover from his days working as a cattle farmer. "Your ears are going to plug up two or three times, guaranteed," he told us as he clattered open the metal platform that would drop us underground. "Swallow, yawn, plug your nose, just try to push back on the drums a little bit."

We piled into a space not much larger than the back of a pickup truck, he slammed the door shut, and we began plummeting. The clanging and rattling of steel against steel made it difficult to hear one another, so we stood mostly in silence as we sank 2,341 feet downward and some 500 feet diagonally north. The noise and speed shook through our legs and chests and eardrums for three long minutes, until the banging gave way to silence and we stepped out into the planet's innards.

The air hung cool against my face with the familiar humidity of a basement, but not a particularly musty one. I had anticipated feeling claustrophobic down there, but I instead felt surprisingly calmed, as if we were nestled within the womb of the earth, inaccessible to any of the troubles whirring up above.

Guided by lights strung on the ten-foot-high ceilings, we approached one of the damp walls to get a closer look. There I found the iconic black and red streaks of banded iron formations that we had seen up above, this time glimmering with splashes of perfectly

AIR

cubed pyrite crystals. While the insides of our bodies stream with iron-rich blood, the insides of our planet, I found, shimmer with iron-rich fool's gold.

The BIFs up in the forest had sat flat as pavement on the surface. But down inside the earth, the magnitude of the ore body came into much sharper focus, and we were glimpsing only a fraction of it. According to some estimations, the formation may carry on for as many as three or four miles deeper underground.

We loaded into an open-air trolley and ventured farther into the mine. About three-quarters of a mile in, Jim stopped the trolley and jumped out to guide us up a spiral staircase into a chamber above, where we were met with the blank stares of mannequins clad in vintage khaki mining gear. With a few clicks of some switches and a warning to cover our ears, he had those mannequins robotically hammer against the wall. The demonstration only lasted a few seconds, but even that felt too long for my ears.

The most common injury among the miners, Jim told us after clicking the hammering off, was hearing loss. "The joke in town," he said, "was that you could walk down the streets of Soudan in the summertime and hear the nightly news without carrying anything with you, because all the windows were open and all the TVs were cranked up so loud."

Aside from their sense of hearing, the miners would have also lacked an understanding of the stories held in the stone that they hammered against. Stanley Tyler and Elso Barghoorn's more elaborate paper on their microfossils hadn't come out by the time the mine closed down, nor had Roscoe's documentation of uranium grains in ancient rivers. The conversation around the GOE, much less the whiffs of oxygen, had yet to take off.

But as much as we have learned since the 1960s, BIFs like these still remain somewhat of an enigma. The deposits in the mine had not remained stagnant in the eons since they formed. Instead, tectonic activity had gathered slabs of crust and concentrated them together like the corners of a blanket crumpled into a ball, tilting the layers on end. As the crust crumpled, cracks formed deep below the surface and created pathways through which fluids could carry dissolved oxygen, rewriting the chemistry of the rocks. The success of oxygen gas as a highly reactive agent has overprinted many traces of its beginnings. The GOE has already erased its own history in a number of places, Woody had told me.

Though the bands looked red to us, Jim confirmed that just beneath the surface they were gun-barrel blue. He recalled a time when he watched a worker drill a hole into one of the mine's walls, and then clear the dust from the hole with a jet of water. "That water was running, out of a blue rock, blood red," Jim told us. "It's instantaneous."

The instantaneous rusting of iron when exposed to the modern atmosphere complicates efforts to verify whiffs of oxygen—some of which have been reported in rocks of this age, 2.7 billion years old. New geochemical imaging technology has helped make progress in this arena, and we would learn more about those methods with Latisha the next day.

As we departed the chamber past the blank stares of the mannequins, Jim told us of a wedding that had once taken place down there. I imagined how those splashes of pyrite crystals must have offered a sort of complimentary, built-in stream of glitter across the walls and ceilings. How that Archean seafloor provided a backdrop, in more ways than one, for how far that couple had come.

✦ ✦ ✦ ✦

THE NEXT DAY WE traveled south to Hibbing, the hometown of Bob Dylan and the world's largest open-pit iron mine. What's now a modest city with wide sidewalks and two-story brick buildings was once the Iron Capital of the World, built on the absence of oxygen in the early atmosphere. It was originally founded two miles north of its current location but was relocated beginning in 1919 when enough ore was discovered underfoot to warrant the upheaval of some 200 buildings and the move south. Funded by tax dollars from the ore, the new high school was built complete with a marble staircase and an auditorium modeled after New York's Capitol Theatre. Today, it still boasts 1,800 velvet seats and ornate chandeliers comprised of thousands of Czech crystals.

Beyond Bob Dylan and what is likely the country's most opulent public school auditorium, Hibbing is also home to the country's largest collection of drill core—cylinders of rock like those Usha and Woody studied in South Africa. Anyone who collects drill core in Minnesota—be it for mineral exploration, engineering purposes, academic research, or any other reason—must turn in a lengthwise slice to the Minnesota Department of Natural Resources, which maintains the Drill Core Library. More than 3 million feet of core have been delivered to this repository since 1972, each sliced down into two-foot segments and meticulously organized in boxes on floor-to-ceiling shelves spread across three Home Depot–sized warehouses.

Those boxes are made available to scientists such as Amy and Latisha and anyone else who might be interested in looking at them for research purposes, or simply out of curiosity. A core collected for mineral exploration one year may, decades later, prove invaluable for academic inquiry. The library hosts local school groups on occasion, and proudly shares its mission of reducing the total number of cores that need to be drilled in the first place in hopes of cutting back on the monetary and environmental costs of collecting them—which sometimes includes clear-cutting a couple of acres of forest and building a road to carry in drilling equipment.

When we met Latisha in a viewing room on the far side of one of the warehouses, she oriented us to the space where she had logged many hours of her career. Wearing hiking boots and black pants tucked into red-rimmed wool socks, she appeared at home in this uninsulated industrial space.

Dozens of boxes of core already sat arranged on a counter that lined the perimeter of the fluorescently lit room. The U.S. Steel Corporation had collected the material not far from the Soudan Iron Mine in the 1960s for mineral exploration; we would repurpose it

64   STRATA

to witness some of the challenges of identifying definitive whiffs of oxygen in the rock record.

"Do you mind if we spray it down?" Amy asked, grabbing a plastic squirt bottle full of water from the counter. With Latisha's go-ahead, she began wetting the stone to help bring out its textures more clearly, revealing surfaces familiar to us from the day before. As I trailed behind her down the countertop, I found that the bottom of the core—the oldest of the rock—contained light-gray basalt, the same as we had seen at our first buggy stop in the woods. Those plain basalts steadily transitioned to a swirlier, mottled mix of basalt and layered iron formation. That mixture finally gave way to hundreds of feet of pure iron formation speckled with shimmering pyrite cubes similar to those we had found in the mine.

Iron formations such as these are notoriously chemically capricious, more so than many other types of rock. Through geologic time, they can change at the whim of the materials that surround them, through processes that geochemists are still working to understand. "The story always seems to be more complex than at face value," Latisha told us.

To cut through the confusion, she and her colleagues have sliced off very thin layers of sample from drill core, embedded these specimens on microscope slides, and are now examining them with intensive optical analyses to determine the order of events that formed their structures. Latisha pays close attention to the shapes and arrangements of the grains in relation to one another and how they intergrow, and then she traces her way back to the mineral phase that would have originally existed in the rock.

"You can start detangling here's what comes first, second, third, fourth," said Latisha, "and tease apart what's being overprinted that way." These are the same types of methods that Woody and his team have used to try to debunk some of the alleged whiffs they have scrutinized.

As we stood huddled around the core, Latisha whipped out a handheld geochemical analyzing tool the size of a toaster oven,

AIR   65

pulled her shoulders up to her ears, and hovered the device above a portion of the core. Within seconds, she was reading off the elemental constituents beneath it from a wallet-sized digital screen on the top of the device. Fe (iron), 75.92 percent; Si (silica), 9.75 percent. The list went on.

Combined with the more detailed microscope work with thin sections that she conducts back at her lab, tools like these are inching her closer to a clearer view of the contents of the rocks and original state they formed within, with a level of precision that has only become possible within the last handful of years.

"I think there is a lot that we will get to detangle in the next coming decade," Latisha said.

In a younger iron formation from 1.9 billion years ago—long after the GOE—Latisha and her team have found microscopically thin concentric layers of rusted iron minerals called hematite trapped between similarly thin silica layers in grains of sand called ooids. As these ooids roll around in coastal areas, they build up layer after layer, similar to snow around a snowball. Every layer had, at some point, been exposed to the open air or open sea, until it became coherent enough for a new layer to form around it.

When a mineral layer gets trapped behind a layer of silica in an ooid, Latisha told us, it becomes sealed off from the environment. Getting the thin, rusted layer of hematite into that grain after the ooid had already formed would be like laying a decal on the outside of your window from the inside of your house, without breaking your window and without opening it up. Pretty implausible.

In this one instance, they can say beyond a reasonable doubt that those nanoscale rusted iron layers are definitive signs of oxygenated shallow ocean waters 1.9 billion years ago. This isn't news—we've long known that at least some oxygen billowed across Earth by then. But the finding is a helpful proof of concept. If Latisha were to identify something like this in cores she is now studying from 2.7 billion years ago, that would be a game-changing clue of a whiff of oxygen hundreds of millions of years before the GOE—an observation more definitive than other poorer-understood geochemical proxies for whiffs.

As oxygen pooled up in the atmosphere during and after the GOE, Latisha reminded us, it would not have accumulated uniformly across the planet all at once, particularly not in the oceans. Storm activity and changes in sea level would have pushed around surface waters and distributed oxygen in different concentrations throughout ocean depths. This inconsistency of oxygen's concentration around the globe—which remains true even in today's fully oxygenated world—further complicates efforts to find a global signal of oxygen through time. Imagine if an alien dropped down to Earth from outer space and landed in Yellowstone National Park today. If they made assumptions about the entire planet based on the rainbow-rimmed hot springs and mud-gushing geysers they found in that corner of Wyoming, well . . . they would have a pretty misguided sense of this place. Geoscientists constantly must consider whether rock samples they are examining are representative of global phenomena, or if they are signals of more anomalous environments like Yellowstone. This is especially true when asking questions of rocks from the Archean, since there are only so many data points to pull from. You can easily find yourself adrift in a sea of false premises.

Later in the day, I asked Latisha a version of the question that I had posed to Woody on that cold morning in Pasadena. Is it worth trying to distill the GOE down to a single trigger? Does that approach make sense?

She exhaled and squinted in the afternoon light, taking a moment to think about it.

"I'm not sure that it does," she finally said. "I think I find the more intriguing answers are the connectedness of the Earth system. The response to a sum of changes."

One thing happened, as Woody had said. But that one thing was tethered to everything else around it.

AIR 67

✦ ✦ ✦ ✦

THOUGH PLENTY OF QUESTIONS remain unanswered about the GOE, each open line of inquiry continues to sharpen our understanding of how the Earth system works and how it withstands change over geologic timescales. By studying this one moment in Earth history, geoscientists have expanded the collection of tools available to untangle other stories in deep time. They've added to their arsenal of proxies. They've grown better at identifying signs of life in the rock record, and at scrutinizing those observations. They've become more adept at detecting red herrings, and have exercised humility when they've been led astray.

The single question of oxygen's rise has frayed into numerous other threads that are collectively drawing us closer to recognizing the real robustness of Earth's stabilizing feedback loops, along with their limitations. We're learning how the Earth system brought us into this moment today, and how we can be more supportive of this system as it carries us into the future.

Geoscientists around the world continue to mull over the unknowns of the GOE, including in the same building at Harvard where Tyler and Barghoorn first huddled over slides of the Gunflint Chert in the 1950s. Andy Knoll, Woody's PhD advisor who himself studied with Elso Barghoorn in the 1970s, continues to work from the top floor of that Harvard building as a professor emeritus. When I met him there in 2022, he was preparing to travel overseas to accept the prestigious Crafoord Prize in Geoscience from the king of Sweden, awarded for his career-long contributions to our understanding of the first 3 billion years of life on Earth.

Knoll spoke with a kind crinkle in his eye and an air of humility as he swiveled in his office chair and clicked through slides of a presentation that he would deliver as part of the ceremony in Sweden. I asked him, as he swiveled back to face me, what he thinks we can say conclusively about the GOE today. What have we learned over the past seventy years since Tyler and Barghoorn published their first paper?

For one, Knoll told me, it looks like it didn't happen on a Thursday. "It was a prolonged event," he said.

It didn't take off 2.4 billion years ago on the dot but, instead, likely unfolded over the span of some 200 million years from about 2.4 and 2.2 billion years ago. We still don't know why, exactly, this gas took so long to accumulate thereafter. Maybe the planet couldn't offer up the nutrients that cyanobacteria needed to proliferate. Maybe oxygen's sinks remained too large for too long, or its toxicity got in the way of more complex organisms evolving to photosynthesize and add to its pool of sources.

One thing Knoll thinks we can say for sure, though, is that the evolution of cyanobacteria alone was also almost certainly not sufficient to drive the event.

"It isn't clear you have this magic pill that you can just drop into the ocean," Knoll said.

Instead, he points to the sum of changes that came about as the insides of the planet matured. He's particularly interested in what Earth's maturation may have done to the availability of the nutrient phosphorus, which cyanobacteria need to photosynthesize. Before stable continents rose up above the ocean, very little phosphorus eroded into the oceans because so few rocks from which it could erode lay exposed to the open air. When the mantle began to mature and cool, newly stable continents sat exposed to winds and rains that sent loads of phosphorous fertilizers into the sea for the very first time. This, in theory, could have allowed cyanobacteria to bloom and proliferate.

Geochemical studies of strata from this time support this idea. It's one of the more than twenty explanations of how a maturing continental crust could have ushered in oxygen's rise long after the very first cells of cyanobacteria emerged. Each of these theories is based on logic pulled from strata, and they each have something to teach

AIR   69

us about how the Earth works, even if some don't turn out to be entirely correct.

Toward the end of our conversation, Knoll mentioned another thread to the narrative that he believes warrants consideration: a series of intense ice ages that arose for the first time around the GOE, and then reappeared again around the same time that oxygen roller-coastered up after the GOE.

Dan Schrag, a geochemist who has studied these intense ice ages and their role in the rise of oxygen, works down the hall from Knoll. When I entered his office, his dog Mickey sniffed my legs and flopped down on the floor next to the sofa where Schrag told me to sit. A past recipient of the MacArthur "genius grant," he has served as the director of Harvard's Center for the Environment and sat on the President's Council of Advisors for Science and Technology from 2009 to 2017. A framed picture of him standing with President Obama and other council members hung on the wall behind me.

As we got to talking about the GOE, I asked Schrag about Knoll's phosphorus angle. He shook his head. "We don't understand that at all," he told me.

When I asked about Woody's idea, that the evolution of cyanobacteria was responsible for the GOE, he shook his head again. "Woody is a lovely guy," he said. But, according to Schrag, he also has it wrong. "There is good evidence now that cyanobacteria evolved long before the Great Oxidation Event."

It was becoming clear that the field of Earth history was built, to a certain extent, on pet theories. To make it in this line of work, you seemed to have to choose your favored plotline and run with it until proven entirely wrong. With so much up for debate and so much room for error, there really wasn't any other way to make progress. Something happened, one thing happened, so everyone can't be right. But nobody was there to witness Earth's antiquity for themselves. All that anyone has is the strength of their preferred theory over somebody else's.

I can't tell you how many times I have heard some iteration of

Schrag's comment about those other takes on the GOE. "ABC is a lovely person," I have heard time and time again, "but they don't know anything about XYZ [insert pet theory, subdiscipline of geology, etc.]." I have learned, over the years, to smile and nod and take it all with a grain of salt. Maybe a hundred years in the future we will have more definitive answers but, for now, it's still very fluid.

Schrag crossed the room to show me a figure on his laptop that illustrated his preferred take on the GOE.

"To me," he said, "this is the story."

He pulled up a graph that illustrated the rise and fall of a series of intense ice ages through geologic time, with oxygen's trajectory overlaid across the graph. Each time one of the ice ages ended, oxygen seemed to have spiked. Those ice ages, Schrag said, were what oxygenated the atmosphere. Full stop.

"The alternative, which some people don't seem troubled by, is to say this is just coincidental," he said, pointing a finger to the screen. "I mean just look at this. That's preposterous."

These ice ages Schrag referred to weren't your typical ones, with frozen water neatly capping the North Pole and the South Pole and leaving lower latitudes alone. These were the most intense ice ages ever to hit Earth. Never before and never since has so much of Earth's surface been covered in ice. With most of oxygen's sinks cryogenically sealed off by this rind of ice, oxygen could have finally accumulated unabated with the help of photosynthesizers growing in and around the ice.

"If we are correct," Schrag said, "there was no inevitability of ending up in a high-oxygen world." Earth could have remained a barren globe full of only rock and water and single-celled beings—no squirrels or shade cast by the branches of jacaranda trees that rustle outside the office windows of Earth historians.

The first of these glaciations, called the Huronian, is the one that Schrag thinks instigated the GOE around 2.4 billion years ago. The next two, called the Sturtian and the Marinoan, came quite a bit later on, spanning between about 717 million and 635 million years ago. Together, these last two comprise a period collectively called the Cryogenian, the age of ice, after which oxygen levels accumulated high enough to support widespread, multicellular life.

If you reach out your arms and imagine Earth's 4.54-billion-year history as a timeline that extends from the tips of your right hand to the tips of your left, with the GOE around your heart, the Cryogenian falls just before your left wrist. The crease in your wrist, the next period after the Cryogenian, marks the Ediacaran, when another major glaciation appears to have swept the planet just before the first widespread evolution of multicellular beings. That leads straight into the Cambrian and its explosion of life that had so befuddled Darwin, when most of the animal groups alive today first came into being.

The stepwise accumulation of oxygen that each of these ice ages may have ushered in could have, in turn, spurred ever larger and ever more complex ways of living on this planet. More complexity meant more organs. More organs meant the capacity to finally eat with mouths and burrow with legs and see with eyes and, long into the future, write with hands.

Getting to the bottom of oxygen's story—to the bottom of our story, and to the bottom of how the Earth works—might mean getting to the bottom of these moments of global ice.

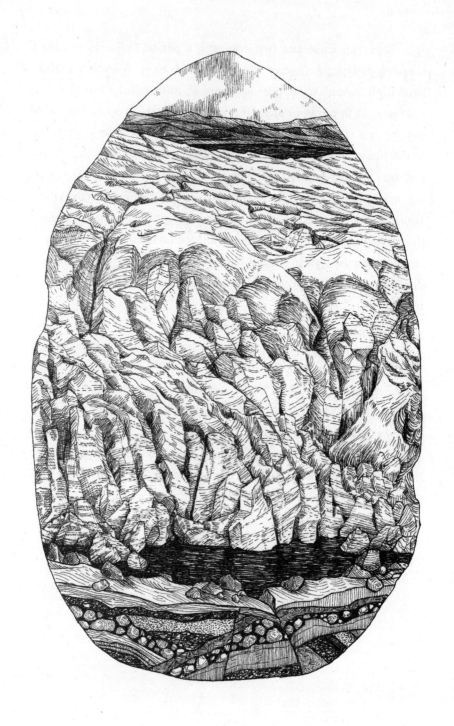

PART II
ICE

The snowflakes land softly at first. One, two, six, nine. They make no noise as they hit bare ground. They drift soundlessly, crystal edges knocking air and light.

Earth spins in and out of darkness in this way, first once, then twice, then on and on. Silence sweeps outward. Snowflakes break and scatter and pile high, then collapse into something stronger than stone. Locked edge to edge, they build glaciers that bulldoze through valleys and screech scars through the land.

As time passes, the air dries up and the snow slows down. But streams still can't bubble. Lakes won't lap. Creek beds whistle in their emptiness and oceans, once bright with light, freeze shut.

Five

WHEN YOU THINK OF ICE, MAYBE YOUR MIND FIRST goes to an object. An ice cube, an icicle, a slushy. Something inert, a thing that can be handled.

At a certain scale, ice does behave this way. But pile up enough of it, enough to create a glacier, and this material takes on its own agency. What once sat blank and unmoving begins to flow and creak and jerk beneath its own weight. Stress and tension crack into crevasses that yawn deep enough to end lives. These forces also spill out beauty, the most brilliant blues that glow like sapphires or hard candies.

A glaciology professor of mine once demonstrated how glaciers flow by holding up a bowl of pancake batter. That batter, he explained to those of us sitting in the dimly lit lecture hall, represented snow. He slowly poured that "snow" out onto the flat surface of the overhead projector. We watched the shadow of the mass, cast against the wall, creep steadily away from the spot where he had poured it. Unable to contain its own weight in the vertical direction, the "snow" slipped out horizontally. It flowed, like a solid mass of liquid.

When an actual glacier flows across an actual landscape, it does not do so as smoothly as batter against glass. It tears apart and reshapes the terrain it travels over. It licks up stones and debris at its base and grinds those bits for miles, streaking and nicking and grooving the land like sandpaper against a length of wood.

A glacier grows through the darkness of winter, gathering more snow that generates more weight and causes the whole thing to heave more heavily outward and downhill. Come springtime, when the sun drifts higher up above the horizon, its warmth melts some of the snow. Gravity sucks that meltwater down beneath the ice, where it gathers and surges as a river, flowing thick with silts and clays and stones. If that fresh water reaches seawater, it will plume out like a bloodstain across the ocean, shifting in size and shape depending on the weather.

We live in a time of pity for glacial ice. It's falling apart, it's melting, it's destabilizing. It's retreating, giving in to warmth like prey to a predator.

These words hold truth, but they also obscure the strength of this substance. Glacial ice is not simply a passive product of the planet's climate; it is an engineer of it, and a ruthless agent of change.

We owe our own existence, in certain ways, to the ice ages of the Cryogenian, those that spanned roughly between 717 million and 635 million years ago that may have incrementally increased oxygen concentrations in the atmosphere and generated a cascade of other changes that rippled across the entire Earth system.

Strata from this period tell us that life as we know it may have bloomed in the wake of Earth's coldest, hardest times. By understanding how the planet rebalanced after this moment of turmoil, we may learn something about the fate of our own crises today.

✦ ✦ ✦ ✦

I SPENT THE SUMMER of 2009 surrounded by ice. I traveled to the Norwegian archipelago of Svalbard alongside two glaciology professors and a handful of other undergraduate geoscience students from the United States who had been selected to participate in a National Science Foundation program designed to help college students learn how to conduct research. Programs like this exist all across the US and elsewhere around the world. We were fortunate enough to have been selected for one some 700 miles from the North Pole, where relatively few humans ever get the chance to visit, let alone work and live.

ICE

Until then, I had known of Svalbard mostly as a setting of Philip Pullman's *His Dark Materials*, a trilogy of fantasy novels I had lapped up as a kid. I didn't realize until late into my teens that Svalbard was an actual place, and I had never expected to find myself on that ragged coastline pursuing what, at that time, felt like a fantasy of a future career path.

Our job was to monitor a collection of coastal glaciers just outside the small town of Ny-Ålesund, where an international research station made logistics easier than they would have been in more remote polar regions. We spent the summer capturing snapshots of how the ice behaved so that, as the global climate continued to warm and unravel in the decades to come, researchers would have points of reference to look back on.

Coastal glaciers like the ones we were studying don't yield to warmth in a linear way, but rather oscillate and negotiate with it circuitously over time. Streams of meltwater, for example, can actually help stabilize the ice they flow out of by releasing loads of sediments that build up in protective mounds in front of the ice. Under the right conditions, these mounds can shield a glacier from the warmth of seawater, like a koozie against a cold beverage.

To color in our understanding of these and other dynamics, I collected dozens of tubes of sediments from the seafloor in front of the glacier while other students collected bits of icebergs and vials of seawater, and made maps of the seafloor. We gathered our samples while working out of a few different motorboats, zipping around like mice at the base of the ice that loomed hundreds of feet overhead.

I had never known frozen water to feel so alive, creature-like in its jerky movements and breathy katabatic winds—gusts of cold, dense air that flowed down the glacier's spine and straight through our hair. That windy wall regularly shed bits of itself into the fjord, with bangs and cracks and swells that we rushed to meet head-on with our boats so as not to be overtaken and thrown in among the polar bears and walruses we were told swam around us. Bearded seals sprawled on iceberg rafts, side-eyeing us as we awkwardly shuffled

around in bulky orange dry suits collecting our samples. We'd sometimes goof around and lose focus, until a bang and a snap would
crack us back to humbled silence and we would, again, reorient our
boats into the swell.

Flocks of puffins trailed our wakes as we motored back to town
each evening to a dining hall full of reindeer stew and other scientists looking to blow off some steam. We lived in red and yellow Scandinavian buildings nestled within the small downtown area
of Ny-Ålesund, where Arctic foxes scurried down gravel roads and
tundra grasses grew surprisingly lush and green, spurred by warmth
exported up from the tropics by the northernmost reaches of the
Gulf Stream. On weekends, we'd file into the one pub in town and
dance alongside researchers from NASA and elsewhere around the
world. Walking back to our bunkhouse in the early morning hours,
the sun still glowing in the sky, we'd sometimes see them dressed in
their orange dry suits, waterskiing the Arctic Ocean under the dusky
midnight light.

Though we were up there to study the climate's unraveling, I was
beginning to see how this work also came with moments of pure
levity, when we could marvel and enjoy the place we were working
to understand and appreciate it for all its quirks and beauty. In those
largely uninhabited, bare and jagged mountains, I felt I could imagine what this planet may have looked like without us, whittled down
to its most basic elements. Air, ice, mud, and heat.

At first we feared the ice. But over time, it became more like a
beloved member of our group and a favorite target of our teasing.
"We're heading to the asshole of the glacier," we'd announce with
smirks over our walkie-talkies to describe our location near the
source of the muddy meltwater stream. We gave nicknames to the
icebergs that carried loads of silt and rocks within them, the ones
that came up from the base of the glacier. They were "dirty bergs,"
and then "dirty, dirty bergs," or "dirty, naughty bergs," and we
snickered along until thunderous cracks silenced us and the glacier
erupted more bits of its face into the fjord.

I had long known climate change was happening, but this was the first time it felt tangible to me. I held pieces of it in my hands, sensed it circling around my neck in the too-warm summer breeze. The glacier had not only formed as a result of a past colder time but also contributed to that coldness by reflecting heat from the sun off Earth's surface back to outer space. With the loss of bright, reflective ice like this came the weakening of the planet's built-in air-conditioning systems that had long helped stabilize the climate.

As we flew south from Ny-Ålesund at the end of the summer research program, I watched the ice disappear beneath low-lying fog. A hollowness in my gut pooled with grief for all that was slipping away—not just the face we had grown to know so well, but the version of the world that could support it. The rupturing of the thermostat that kept us intact.

What I didn't know then was that the rocks on the land that surrounded our boats held stories of other periods of glacial growth and decay—stories that could help us get a better grip on our current moment of global change.

✦ ✦ ✦ ✦

DECADES BEFORE I ARRIVED on Svalbard, a young British geologist named Brian Harland traversed its valleys dozens of times over and scratched his head at the story that seemed to unfold in the maps that he drew from those strata. A photo captured during the first of more than forty expeditions he would take shows a scruffy-haired twenty-one-year-old with a wide, toothy grin and high cheekbones that frame earnest eyes. With shoulders pushed back and head leaned toward the camera, he looks totally in his element amid those mountains and fjords. That was 1938. The outbreak of World War II the following year would prevent him from returning for some time. When he finally did make his way back a decade later, he did so as a professor of geology at the University of Cambridge, bringing a cadre of students along with him.

As Harland and his students made their way across the icy terrain, they pulled together evidence of another icy time that had taken place on that land hundreds of millions of years earlier. They found a few telltale signs of glacial activity in the ancient strata that they recognized as such thanks to their understanding of glacial activity in the modern world. For example, whereas rivers of water struggle to move debris larger than dinner plates in diameter, rivers of ice pluck up stones the size of golf carts and push them along like dust in a dustpan. These boulders get mixed in with much finer sand and clay to form what's called a till. The randomness of glacial till distinguishes these deposits from those laid down by more delicate and organized movements, such as the flow of water through streams or out in the open ocean. Harland and his students recognized glacial till because it lay in plain sight among modern glaciers.

Unsorted mixes of sediments can also accumulate from other sorts of geologic activity, like landslides, and so they looked for a suite of other clues in the strata to weed out alternatives and affirm they had, indeed, found evidence of ancient ice. When glaciers move across the ground, they drag rubble beneath them, carving shallow, parallel grooves called striations through the bedrock. When icebergs drift out to sea, they melt out rubble that drops to the seafloor as pieces of debris called glacial dropstones. Because dropstones fall from above, they can pierce and depress the fine silty layers of the seafloor and cause those layers to warp around them, like a sofa cushion hugging a body.

Of all the visual clues used to identify glacial deposits in the rock record, dropstones often provide the most convincing evidence of ice. Few other forces can send rocks dropping from above in a body of water. Uprooted trees can carry stones in their roots and release them as they get caught up in a current, but the strata Harland was studying had formed hundreds of millions of years before trees had even evolved.

Check, check, and check. All of these signs of glaciers present in the modern sediments of Svalbard also appeared in the ancient strata.

ICE

Harland understood that ice ages come and go, so evidence of ancient ice probably didn't surprise him very much. But what *did* surprise him was what he and his students found interspersed among the glacial deposits: thick layers of yellow and gray carbonate rock.

Today, carbonates form the foundations of Australia's Great Barrier Reef, the Bahama Banks, and other such tropical locations. They typically form only in the world's warmest waters, which rush around the equator. If their interpretations were correct, Harland and his students had found evidence not only that Svalbard had once been closer to the equator but that, at some point while it was there, it had also been encased in ice.

This, Harland realized, was odd. In today's world, glacial ice can persist in equatorial regions such as the Andes and Indonesia, but only on mountainous slopes several miles above sea level. If ice had formed all the way at sea level in the warmest part of the world—the part with the most direct sunlight, around the equator—then ice probably would have smothered much of the rest of the planet as well. If that were the case, this would mark the most extreme ice age known in all of Earth's history. At no other time had glacial ice made it to sea level that far from the poles. The most recent ice age that peaked some 20,000 years ago covered only about 8 percent of the planet; Harland's ice age looked like it covered closer to 100 percent (though he wouldn't come to this conclusion until years after his initial observations).

From outer space, this cryogenic version of Earth would have glistened white and bright. Beneath the surface of the frozen ocean, dark and briny currents would have stirred continuously from geothermal heat billowing out of cracks in the seafloor. Up above, winds would have howled around the equator in strong gusts, and tiny ice crystals would have settled out of air that remained below freezing everywhere. Life persisted through all this, but it couldn't have been easy. Algae, bacteria, and simple protozoa would have squirmed around on top of or under the ice, and other specks of life would have swarmed alongside hydrothermal vents on the seafloor.

This all may seem a bit fantastical, but Harland couldn't see any other way to interpret the Svalbard strata. When he and colleagues conducted follow-up studies of magnetic minerals within these deposits, they found that those grains—which behave like fossilized compass needles—pointed to a near equatorial origin for the rocks as well.

"It is concluded," he wrote in a 1964 paper documenting his unusual finds, "that an ice age was sufficiently extreme to form marine tillites in the tropics."

That is, glacial tills. In the oceans. In the tropics.

Energized by these findings, Harland presented his theory to colleagues. But it was a hard sell. Rather than accept his seemingly fantastical ideas, some wove together alternative theories. They suggested that those sediments came not from ice but instead from something more mundane, like a series of widespread and robust mudflows.

But Harland persisted. Mudflows, he argued, don't drop uniformly over thousands of square miles, and they don't leave behind dropstones. Further to his point, he wasn't the only geologist finding evidence of ancient glaciers in rocks of roughly this age. As early as 1891, a Norwegian geologist reported deposits of a similar age along the shores of a fjord in northeastern Norway that had striations— those long grooves that form when glacial ice sandpapers the ground beneath it. Over the intervening decades, Harland and others identified some two dozen localities of similar randomly sorted, striated, dropstone-covered rocks across the world in deposits of a similar age, from East Asia to Western Africa to South Australia and Greenland— many with the same curious layers of carbonate capping them off.

Individually, the glacial interpretations of these rocks didn't raise many concerns. But when they were taken all together, and the warm-water carbonates were thrown in, the story became harder to stomach.

These glacial tills weren't simply a blip here and a blop there. They were substantial. They measured between ten to one hundred meters thick, sometimes with boulders as large as baby rhinos. To

ICE 83

Harland's mind, these were all too expansive and widespread to have come from mudflows. The suggestion that they had was more outlandish to him than the possibility of ice at the tropics.

The mudflow rebuttals drifted away over time, but many of his colleagues still couldn't accept the idea of ice covering the turquoise waters and soft sand beaches that they thought of as the tropics. They couldn't imagine a version of Earth that could ever get that cold, nor could they see how the planet would spin out of such a deep chill. They couldn't, it seemed, grapple with the possibility that Earth was once much different than it is today.

These colleagues continued to come up with more alternatives. Some called for wandering magnetic poles that would have shuffled across the planet and brought cold snaps directly to continents, like a waiter serving drinks around a dining room. But that hypothesis, Harland wrote in a bristly 1964 *Scientific American* article, was not supported by any concrete evidence. "Indeed," he and his coauthor, Martin Rudwick, lamented, "it seems to have been suggested only in order to avoid postulating the presence of ice in the tropics."

Harland and Rudwick were not willing to be so avoidant. "Failing any other explanation," they wrote, "we are prepared to accept that [ice] did exist there."

That those brilliant turquoise waves had flattened and hardened into muted whites and grays. That smells of salt gave way to ice's absent scent.

The desperation with which Harland worked to understand this ice age stemmed not only from its strangeness but, perhaps even more so, from its timing in relation to another major event in Earth history. Relatively soon after the last of the ice deposits drift away, a burst of fossils appears around the world. This marked the beginning of the Cambrian period, when the majority of animal groups alive today first appear in the rock record, in that explosion of life that had so befuddled Darwin.

"It can hardly be mere coincidence," Harland and Rudwick wrote in their 1964 article, "that a geological event of such intensity was

84   STRATA

followed, after a relatively short interval, by a biological event of equally striking character."

To their minds, the ice age could have instigated the explosion of life that followed. Getting to the bottom of this global cooling event, which they named the Great Infra-Cambrian Ice Age, might reveal how the planet produced the complex multicellular beings that would one day evolve into us.

✦ ✦ ✦ ✦

AROUND THE SAME TIME that Harland began advocating for this worldwide ice age, a Russian geophysicist named Mikhail Budyko was trying to understand how ice at the poles influenced global climate, but from a more theoretical standpoint. He became interested in this, in part, out of curiosity in the ways that the growth and decay of Arctic ice might affect aridity and, in turn, agriculture across what was then the Soviet Union. He wanted to figure out just how tenuous the global climate system was, and how beholden different facets of the climate were to the extent of ice at the poles.

These concerns boiled down to a simple concept: dark surfaces absorb more heat from the sun than lighter surfaces do. We can feel this within our own bodies when we wear dark clothing or duck into a car with dark seats on a hot summer day. When we put on a lighter-colored shirt or duck into a lighter-seated car, we feel cooler.

What proves true on a human scale in this case also proves true on a global scale. Take a dark, open ocean and cover it with white reflective ice, and the planet will cool. Its albedo, or reflectiveness, will increase. Under the right circumstances, ice will beget more ice in a cooling loop that will continue on and on until something else within the Earth system—say, a series of intense volcanic eruptions— triggers a warming response that balances things back out. (Today, the melting of ice at the poles is revealing more dark, heat-absorbent open ocean, begetting more melting in a warming loop running in reverse of this cooling one—although cloud cover above the ocean

ICE 85

can reflect some of that radiation back out to space, further complicating the feedback loop.)

To test the limits of these dynamics on a global scale, Budyko created a mathematical model that loosely replicated Earth's climate system. To simulate change within that system, he varied the solar constant—the amount of energy that the sun sends to Earth across a given area. This wasn't so much an effort to determine how the sun's strength affects global climate but rather how the planet responds to any nudge in global temperature. It was a stand-in for a whole array of conditions that might shift Earth's thermostat.

When he dimmed the power of the sun by 1.5 percent of its modern brightness, he produced the conditions of an ice age similar to the one that peaked some 20,000 years ago. Glaciers covered part of the planet, but hovered stably around high latitudes. Then he dimmed the sun just one-tenth of a percent more, and everything went white.

That nudge had been enough to push the planet from a partial chill to a chill so deep it sent ice racing to the equator. Once the ice reached past 50 degrees latitude in his model, there was no turning back. Ice had begotten more ice until the entire planet had become encased in it. (More recent research suggests that number is closer to 30 degrees latitude.)

While his findings were intriguing in theory, Budyko and his colleagues wrote them off as implausible in reality. They couldn't fathom that such a White Earth Disaster, as some called it, could take place here, because they couldn't see how Earth would ever spin out of it.

Meanwhile, isolated off in a separate sphere of geology, Harland was massaging his temples and pointing to evidence of just such a disaster on nearly every continent around the world.

Six

THE FACT THAT BRIAN HARLAND COULDN'T YET explain how this ice age may have started or ended wasn't reason to give up on his theory. After all, in the mid-twentieth century, when he first started finding these deposits, the planet's inner and outer workings still remained largely unknown. Nobody could explain exactly how the continents moved, how oceans formed, why earthquakes shivered through the crust, or how pieces of the crust changed over time. There were theories forming, but much remained up for debate when Harland began mapping those Svalbard rocks in the 1940s. He had been an early proponent of the theory of continental drift (proposed in 1912 and a precursor to the theory of plate tectonics), and was willing to accept that his Arctic rocks had somehow originated in more southern locales. But he couldn't have any hope of knowing exactly why or how they had moved across the surface of the planet until more than two decades later, when geologists finally started nailing down the finer details of plate tectonics.

Trying to make sense of global climate without understanding plate tectonics is like trying to ride a bike without a chain. The continents and oceans and atmosphere all run the same machine. Take one out and you will spin your wheels to nowhere. We now know that when two tectonic plates collide and buckle the crust to form new mountain belts, freshly exposed rock reacts with carbon dioxide in rainwater and, in so doing, can cool the atmosphere and the oceans. When plates slide over hot spots, volcanoes erupt and spew out greenhouse gases that, in turn, can warm the atmosphere and the oceans. These are just some of the many feedback loops that power the Earth system and its climate. Without the deep understanding of all of these yo-yoing interactions, Harland would have had no hope of making sense of any ice age, let alone a global one.

Eventually, though, enough pieces of this puzzle fell into place to make sense of plate tectonics and, over time, those global glacial deposits. For this, we have American oceanographer Marie Tharp largely to thank.

As a woman, Tharp was a rarity in the geosciences when she took her position as a cartographer at Columbia University in 1948. Until the late 1960s, she and many of her female colleagues were prohibited from joining their male colleagues out in the field. Instead, Tharp spent her days at her laboratory desk, sketching out seafloor maps to make sense of the sonar readings that her male colleagues collected out at sea.

As she sketched out those millions of data points in pen and ink, Tharp discovered the sprawling contours of the Mid-Atlantic Ridge—a massive underwater mountain range that stretches more than 10,000 miles long and rises nearly as high above the seabed as the Rocky Mountains do above North America. The scientific community had long known that some sort of ridge existed in the middle of the Atlantic, thanks to echo soundings collected as early as the mid-nineteenth century, but only once Tharp drew out her map could they see this ridge in its full dimensions.

In her profiles of the seafloor, Tharp also discovered the dramatic relief of a gaping valley that dipped wider and deeper than the Grand Canyon and that sliced straight through the middle of the Mid-Atlantic Ridge. Here, Tharp had discovered what we now know as the Atlantic's seafloor spreading center—where the North American and South American Plates pull away from the Eurasian and African Plates, steadily widening the basin of the Atlantic at the rate that our fingernails grow each year. As tectonic plates rift apart from one another in places like this, they thin the seafloor and allow new magma to rise up from the mantle. To make room for this new crustal material, the far ends of the growing plates travel down beneath neighboring plates, in a conveyor belt of movement that allows continents to drift across the globe. Though German geoscientist Alfred Wegener had argued for something along these lines in his theory of continental drift decades earlier, in the early twentieth century, only once Tharp's map revealed the rift valley of the Mid-Atlantic Ridge could the scientific community make sense of how this drifting might actually take place.

Tharp's discovery of the rift valley was groundbreaking, but it was initially dismissed by her male colleagues as "girl talk." Over the following months, though, those colleagues not only grew to accept her interpretations as valid but also took credit for them as their own. Tharp was not listed as an author on the 1956 publication that shared her discovery with the scientific world.

"I worked in the background for most of my career as a scientist," Tharp later wrote, "but I have absolutely no resentments. I thought I was lucky to have a job that was so interesting."

While the geosciences have become more accessible to all genders in the United States since Tharp's time, non-male geoscientists still face obstacles that keep many from rising up the ranks in their field. A study published by the American Geosciences Institute found that though women made up 39 percent of all geoscience graduate students in the US in 2017, they made up only a quarter of the postgraduate workforce. Another study that explored why women don't stay in the geosciences found that male-dominated work cultures, a lack of female role models, and experiences of sexism all contributed to their exodus. Indeed, all these rationales resonated with my own decision to choose a different path even before pursuing a graduate degree in the geosciences.

In their book *Forces of Nature: The Women Who Changed Science*, authors Leila McNeill and Anna Reser describe how the relegation of women to the margins of science dates back centuries. I had long known this to be true, but what I hadn't realized until I read their book a decade after I became a science writer was this: "Women turned to science writing and popularization in order to continue to feed their interest in the study of nature, while the male scientists, working in the protected spaces of scientific institutions, regularly used women's scientific work without ever giving them credit."

As Tharp pulled together more sonar readings from around the world over the following years, she and her colleagues found more than 40,000 miles of valleys and ridges that, together, laced the planet like seams on a baseball. In 1977, she copublished the World Ocean Floor Map, the very first depiction of the global seafloor in all its textured relief, with the outline of the planet's tectonic plates revealed in sharper detail than ever before. When overlaid with seismograph readings, the map indicated that all of the ocean's major earthquakes took place within the confines of the rift valleys that Tharp had discovered—further illustrating how these valleys seethed with tectonic activity.

Tharp's largely unsung contributions to geology not only presented the scientific community with a clearer understanding of plate tectonics; they also offered a healthy dose of humility. A reminder of how much there was still left to learn about how the Earth worked.

✦ ✦ ✦ ✦

IT WOULD TAKE A geologist who came of age during the time of Tharp's discoveries, who felt the very foundations of geology metamorphose with this new understanding of Earth's innards, to make strides in understanding the radical idea of the global ice age that Harland advocated for.

Enter Paul Hoffman. This Canadian geologist earned his undergraduate diploma in 1964, right in the midst of the tectonic revolution that taught him to take what he was told about how Earth worked with a grain of salt.

"One of the most important lessons I learned by coming of age during the plate tectonic revolution," he told me over a phone call, "was to not pay too close attention to advice offered by more senior people." Hoffman's mentors told him not to study tectonics because tectonics were hopeless, he said. "It was impossible to understand the basic causes of mountains and volcanoes and all that because it was beyond our possibility of knowing."

He ignored that advice and went on to spend twenty-five years of his career working for the Geological Survey of Canada, occasionally taking stomach-churning dips in Stu Roscoe's plane, and mapping out the forms and structures of ancient rocks to try to make sense of the tectonic histories embedded within them. He loved the physical component of the work and the opportunity to think about the big picture of the planet. He thought he might stay in that position forever.

But, over time, the Geological Survey began to shift its priorities away from academic research and more toward regulatory oversight and resource extraction. Not one to keep opinions to himself, he openly voiced his dismay at the drift away from scientific inquiry. Soon after, he stopped receiving funding for his fieldwork. He had no choice but to find a new line of research in a different region of the world. After a brief stint at the University of Victoria, he landed a job at Harvard, in 1994. (He was the Harvard geologist who told me that Woody Fischer felt like an addition to the faculty when he arrived on campus in the early 2000s.)

After leaving the Geological Survey, Hoffman went to work sifting through bedrock geology maps in search of a new focal point for his research. The Republic of Namibia in southwestern Africa stood out to him as one promising option for its very old rocks, deserts that kept those rocks vegetation-free and visible, and political stability that would allow him to return year after year. With funds granted by Harvard to continue on with research he had begun at the University of Victoria, he bought himself a truck in Namibia and set off to inspect the strata there.

By that time, he had heard of the Great Infra-Cambrian Ice Age, though he knew it by a different name. While the idea of a global cold snap may get your heart pumping, the phrase "Great Infra-Cambrian Ice Age" almost certainly will not. In 1992, a researcher at Caltech named Joe Kirschvink published a paper that cleverly re-coined this period as "Snowball Earth" to signal the real magnitude of this ice age; to indicate that this was a time in Earth's past when the entire planet would have looked, from outer space, like a giant snowball.

Kirschvink had shared what he knew about the theory with Hoffman several years earlier, while they were both in Washington, DC, for a geology conference. They were sweatily waiting for a bus on a hot summer day, and the idea of a global ice age sounded "delicious" to Paul. But beyond that, it still lacked some credibility and he didn't think he had anything in particular to contribute.

As Hoffman whipped out his hand lens and wielded his rock hammer in Namibia a few years later, though, he found strata that looked strikingly similar to those that Brian Harland had described from Svalbard. Those same unsorted mixes of boulders, pebbles, and sand that were indicative of glacial activity, all closely associated with thick carbonate deposits indicative of the tropics. He had heard of these deposits before arriving, and the paradox they seemed to represent had intrigued him. But he wasn't thinking seriously about Snowball Earth at that time. Like many of his colleagues, he remained pretty skeptical of the idea.

Paul brought samples of the carbonate back to his lab at Harvard, where he worked with others in his research group to explore what their chemistry might reveal about the world those strata formed within. The team specifically measured the ratio of carbon isotopes—the different flavors of carbon with varying numbers of neutral particles called neutrons. The suite of carbon isotopes found in nature includes carbon-12 (6 protons and 6 neutrons), carbon-13 (6 protons and 7 neutrons) and, less frequently, carbon-14 (6 protons and 8 neutrons).

As marine organisms like phytoplankton weave carbon into their cells during photosynthesis, their cellular machinery preferentially pulls in carbon-12, presumably because it is the lightest-weight isotope and therefore easiest to manage. The heavier carbon-13 gets left behind in seawater where, under the right conditions, it becomes incorporated into carbonate rocks.

This is all grossly simplified; isotope geochemistry is far more complex. Even so, geologists have used this basic premise to try to draw conclusions about the global environment from changes in carbonate chemistry in the rock record. For example, carbonate rocks without much carbon-12 in them might represent times when marine life was abundant and hoarding all the lightweight isotopes in the oceans, whereas carbonate rocks with more carbon-12 might represent times when marine life was less abundant, thereby leaving behind more of the lightweight isotope in seawater to get incorporated into rocks.

When members of Hoffman's research group loaded their carbonate samples into their laboratory equipment to measure the isotopic contents of the rock, they were surprised to find very lightweight carbon isotope readings. The readings were so lightweight, they suggested that the oceans were all but absent of life. Less experienced in carbonate geochemistry than field geology, Hoffman wasn't quite sure what to make of the numbers, but he continued gathering samples from Namibia for the next several years. He was convinced that he was working within glacial deposits, but was uncertain what the geochemical readings meant or what, if anything, they had to do with the theory of Snowball Earth.

He shared his befuddlement with a colleague visiting Harvard as a guest lecturer in November 1997, and she pointed him to a paper on seawater carbonate chemistry by Lee Kump (one of the proponents of whiffs of oxygen). A night owl, Hoffman returned to his office after dinner to read the paper she had recommended. "It was the eleventh hour of the eleventh day of the eleventh month," he recalls with the intensity of someone whose life was about to change.

ICE 93

While reading the paper and learning more about the intricacies of carbonate geochemistry, Paul had an epiphany: one way to pause life within the oceans and get the lightweight values he was seeing in his carbonate samples could be to seal off those oceans with a thick lid of something. Snowball Earth came sliding into focus as at least one viable explanation. The isotope readings alone couldn't explain how the planet fell into or out of this deep chill, but they were enough to drive Hoffman into action.

For the first time, he felt that he had something to contribute to Snowball Earth. He had been looking for a life in geology after his work on plate tectonics in Canada, and had found it in this global ice age.

He tempered his initial excitement with a healthy dose of skepticism, asking himself every day whether it could really be true that the planet had been covered entirely in ice for millions of years at a time. And then he would talk himself down from these doubts. "I would say to myself, 'Well, just settle down, Hoffman,'" he told me. "If you look at it as a cold, hard scientist, it explains so many things for which there's no other explanation."

Dan Schrag, who worked with Hoffman at Harvard at the time, provided a sounding board for his musings. Together, they drafted a paper along with a couple of other colleagues and published their arguments for Snowball Earth in the journal *Science* in 1998. As was the case with Harland's early papers, theirs wasn't warmly received by many of their colleagues. But, by then, Hoffman was too enthralled by the possibility of the theory being correct to let it go.

Although the isotope findings later turned out to be less straightforward than they originally thought, they pushed Hoffman past his initial hesitancy to become involved in Snowball Earth research. He has since spent more hours consumed by this period than any other human on the planet—though many other humans have since jumped on the bandwagon along with him. He has lived more than fifty months out in the field in Namibia and continued traveling to work there even into his late seventies. He has dug into the problem from many different angles and has collaborated with colleagues from around the world across many different fields of science, from climate modeling to physical oceanography to evolutionary biology to microbiology.

All along, he has faced pushback precisely as Harland had. And, like Harland, he has used that pushback not as a reason to back down but as a reason to dig in his heels and try to strengthen his case.

"It's just such a large and fascinating problem," he told me. "It's hard to pull myself away."

After all, he noted, people struggled to believe that most of northern North America was covered by an ice sheet some 20,000 years ago when theories of the most recent ice age surfaced in the 1800s. That was as hard for nineteenth-century humans to accept as Snowball Earth had been for twentieth-century humans. Incredulity is not a reason to give up.

Now in his eighties, with scraggly white hair that matches a trim-cut beard, he speaks with a voice hoarsened by decades of making his points loudly known. His eyes soften into a warm smile as he cracks the occasional wry joke, but otherwise they remain steady and focused as he talks about his research.

New ideas are like small children, he likes to say. You can't know how they'll grow and mature unless you give them time to do so. "Very often the problem is not that they are wrong," he told me, "but they are incomplete."

✦ ✦ ✦ ✦

WITH YEARS OF OBSERVATIONS from the field, and with the help of climate modelers who have filled in details that the rocks alone cannot, the picture of Snowball Earth and its role in the spread of multicellular life thereafter has slowly come into sharper focus. Plenty remains uncertain and up for debate, but some of the haze of Harland's time has finally cleared and given way to a theory that looks something like this:

ICE   95

Within thousands of years after the planet first fell into a runaway cooling loop, the oceans grew a rind of ice so thick that it flowed like glaciers do on land. Underneath this global sea glacier, powerful submarine currents ribboned and rushed, propelled by heat emanating up out of vents in the seafloor. Bacteria and other simple lifeforms subsisted on geochemical trickles of energy bubbling out of these vents.

Atop the ice, winds howled across an otherwise quiet surface, whipping air that grew too cold and dry to harbor humidity. Temperatures hovered below freezing everywhere, dipping below negative 20 degrees Celsius in the tropics.

Snow and ice covered much of the continents, but some patches of ground remained exposed to the air, as is true in parts of Antarctica's McMurdo Dry Valleys today. When winds whipped over this bare land, they lifted dust and spread it across the surrounding ice, darkening otherwise bright, white surfaces. Where those surfaces darkened enough, the dust melted tiny holes in the ice. Fungi, bacteria, algae, and other simple lifeforms floated and squirmed in these oases, fueled by nutrients brought by the dust. That pigmented life, in turn, further darkened and expanded those little bodies of water. Holes the size of marbles grew to the size of golf balls, to the size of basketballs and even larger. The ice surface became a frozen Swiss cheese, pockmarked all over the place with dust-borne puddles (such puddles appear on ice surfaces around the world today). This band of dirty ice straddled the equator and was, in Hoffman's opinion, the Snowball's most striking feature.

In some places, the holes merged and coalesced into hip-deep ponds that could have persisted year-round at the equator. These unfrozen habitats helped life survive through the duration of the Cryogenian.

Slivers of the sea surface may have remained unfrozen as well, in places where geothermal heat rose from the seafloor or where sea ice cracked and rubbed against land. If even only 1 percent of the ocean's surface had remained open ocean—a small enough fraction to keep the snowball intact—hundreds of thousands of square kilometers would have still remained available for lifeforms to persist and evolve.

And evolve they did. Some left behind microscopic fats called steranes that geochemists have since resurrected from Cryogenian rocks to get a sense of what lived through this time. Evolutionary biologists have also developed so-called molecular clocks to back-calculate what may have been alive during this period. They do so by modeling how different genes mutate and estimating when different modern groups of organisms first split away from each other.

Together, steranes and molecular clocks suggest that the planet's very earliest animals may have arisen sometime between about 713 and 635 million years ago, in the form of sponges. From these very simple filter-feeding bodies, other animals evolved.

But how might the conditions of the Cryogenian have spurred this evolution?

"The easiest way to get new species is to geographically isolate them," says Phoebe Cohen, a paleobiologist who studied with Andy Knoll as a graduate student at Harvard and now works at Williams College.

A species, Cohen reminded me, is a group of organisms that shares genes and can successfully reproduce with each other. If you split up a population of a species with some sort of geographic barrier—say, a sheet of ice—and prevent those isolated populations from reproducing and sharing genes, then they may eventually go their own evolutionary ways and develop into different species. Even if that geographic barrier eventually melts away, the organisms will have diverged too much to successfully share DNA.

"Snowball Earth in many, many ways, must have been a great way to do that," Cohen says. The freshwater puddles and ice-free terrestrial environments scattered across the planet would have left Earth's habitats isolated and ripe for speciation.

ICE
97

Still, it is all but impossible to pinpoint this as the sole or main driver of the diversification of life that followed the Cryogenian, since there's only so much stratigraphic material to pull from across this interval. Cohen isn't even particularly comfortable drawing the conclusion that Cryogenian conditions directly led to the Cambrian explosion of life, given that some 90 million years spanned the end of the one and the beginning of the other. Though that is not very long in the scheme of Earth's entire history, it might be too long to draw a direct cause-and-effect relationship. By way of comparison, brontosauruses roamed the earth some 90 million years ago. "Snowball Earth was obviously a critical event in Earth history and had a big impact on life, but it precedes the Cambrian radiation by tens of millions of years," Cohen says. "We have to do our best to try to figure out what we think happened based on our understanding of how evolution works in the modern day."

Even given the limitations of the rock record, those working in this line of research still have their preferred pet theories.

Paul Hoffman agrees that geographic isolation during the Cryogenian could have forced living things to diversify and innovate new ways of living that ultimately could have driven the explosion of complexity that arose later on.

Dan Schrag disagrees.

"I love Paul, but I think he is not seeing the big picture," he told me. "I think that is kind of army-wavy."

Instead, Schrag thinks that the role that Snowball Earth played in the rise of oxygen—by sealing off oxygen's sinks and allowing it to accumulate in the atmosphere—led to the rise of complex life thereafter.

"Oxygen opens the door and lets things get big," Schrag told me from his office at Harvard. It helped break down food and allowed metabolisms to become more ambitious and complex. It paved the way for the first predator-prey relationships that, in turn, paved the way for shells, spikes, burrows—means of defense that proliferated during the Cambrian with the help of the oxygen brought on by the Cryogenian.

"To me, that is a really compelling idea," Schrag said.

Cohen isn't as compelled. Yes, organisms needed a certain amount of oxygen to grow larger and more complex, but some geochemical proxies suggest that oxygen levels were already plenty high enough during Snowball Earth to allow for some degree of diversification of complex life—yet it doesn't appear to have taken off until later on. The jury is out on oxygen as the driver of diversification, at least in Cohen's mind.

Still another theory, which Cohen does find rather compelling, suggests that the high viscosity of Snowball Earth oceans led to the explosion of complex life. As water cools, it becomes more viscous and more difficult to travel through, at least for the smallest of creatures. Lifeforms may have evolved new ingenuities to work together to travel through those cold, thick waters, and these innovations may have led to the proliferation of multicellularity.

Perhaps. We can't know for sure, at least not yet.

"It is impossible to say one thing is more important than another because we don't really know," Cohen told me. "This is all based on what we think was happening, but we have so little direct evidence of what happened."

The problem, to borrow Hoffman's motto, may not be that these ideas are wrong, but that they are incomplete.

In a 2017 paper that Hoffman and Cohen coauthored with some two dozen other colleagues studying the Cryogenian, they nod to their persistent need to remain wide-eyed and skeptical of their own ideas, particularly as they rely heavily on computer models to flesh out those ideas.

"Like geology itself, models are a source of astonishment, wonder, and inspiration," they write. "Like geological observations, they are incomplete and subject to improvement."

How to improve the models and geological observations? Spend more time with the strata.

ICE

✦ ✦ ✦ ✦

A WEEK AFTER GRADUATING from college in 2010, I stepped off a plane in South Australia and into the world of ancient ice. Through a bit of luck and some pesky persistence, I had weaseled my way into a job working as a field assistant for a group of researchers studying Snowball Earth in the Flinders Ranges, a collection of mountains that rise like red shark fins out of the Outback. I had wanted to test the waters of academia as a career path after graduating, and so I had sent a handful of emails out to geologists whose work interested me, asking if they were looking to hire an assistant for the following year. Most politely declined, but one agreed to chat more—a stratigrapher and Earth historian at Princeton University named Adam Maloof. He had been a graduate student of Paul Hoffman's at Harvard in the late 1990s and early 2000s and, under his tutelage, had become a Snowball Earth expert in his own right.

As a field assistant for Adam's group, I would help carry loads of samples in my backpack, scratch down observations in my water-proof field notebook, and contribute to the interpretation of the strata when I could. We would be collecting stratigraphic logs, centimeter-by-centimeter observations from the base of rock formations to the top. Individually, these barcodes of the earth wouldn't reveal much. But many barcodes collected across a region would provide a multidimensional view of the landscape and how it changed through time.

In his email offering me this position, Adam emphasized with italics and underlines and asterisks that this field season would be a long one. They had a lot of samples that they wanted to collect, across dozens of field sites, and all of that sampling work took time. Geologists typically spend a few weeks or maybe a month in the field, and then return back to their lives at home. If I took the position, I would jump between different regions and projects for three and a half months. We'd live out of tents, rarely shower, and go weeks without seeing humans other than those within the research group. He encouraged me to think hard about whether this was something I was physically and emotionally prepared for.

But I was twenty-two years old, in need of a job, and enamored with the idea of living out of a tent in the Outback. This was not something I would think twice about. A few months later I was walking the floors of ancient fjords, rock hammer hung from my hip, surrounded by emus and kangaroos.

The motivation to study Snowball Earth had, by 2010, expanded to encompass questions about modern climate change and the future stability of the planet. By understanding what has broken the climate system in the past, Adam and his colleagues reasoned, we might learn something about how our current climate crisis will unfold today. How different parts of the Earth system might kick into gear to balance things back out.

During previous field seasons in the Flinders, Adam's research team had discovered a collection of strange shapes embedded within one interval of 650-million-year-old strata. When they brought samples back to the lab to study them more closely, they found that the shapes contained tiny canals that looked like they could have belonged to a filter-feeder, like a sponge. If they were correct, they would have found the very oldest animal body fossils known of on the planet. We would have a chance to visit those possible sponges and continue to try to make sense of how Snowball Earth may have influenced the evolution of complex life during and after the Cryogenian.

By this time, those studying the Cryogenian had come to recognize it as not one but two separate glaciations—the Sturtian and the Marinoan. The Gaskiers glaciation, the next and final widespread ice age that fell some 579 million years ago, ended only 1 million years before the world's first widespread multicellular lifeforms appear in the global fossil record. The Gaskiers had striking similarities to the glaciations of the Cryogenian, but whether it was a full-fledged Snowball Earth remained an open question. It also wasn't clear why

ICE   101

the Sturtian lasted so much longer than the Marinoan, and why there were nonglacial interludes between these glaciations.

Another point of contention was the extent to which the entire planet froze over during these events. Where Woody Fischer has had to contend with proponents of "whiffs" of oxygen leading up to the Great Oxygenation Event, Hoffman and others in his camp have had to contend with those who call not for a Snowball Earth but something softer. A Slushball Earth.

The term Slushball Earth means different things to different people, and is not a phrase that Hoffman particularly likes—both because it pokes fun at the science and because it's not descriptive of the version of the planet it describes. "It doesn't have anything to do with slush," he says.

The term originally referred to the results of a climate model published in 2000 that generated a computer-simulated version of Earth in which all the continents froze over but the tropics remained ice-free. The researchers who produced the model presented it as a theoretical alternative to the full-fledged Snowball Earth that Hoffman had been advocating for. It appealed to paleontologists struggling to rationalize how life persisted through a total global freeze-over. But Hoffman was quick to point out the model's shortcomings—including that the event it produced did not last nearly as long as the geologic record suggested the Snowball Earth events had lasted.

Other climate modelers have since generated another alternative to the full-fledged Snowball Earth that Hoffman believes is based on sounder science, but that still has limitations. Called the Waterbelt, it's a version of Earth that is mostly frozen over, but has a very thin band of open ocean around the equator. To Hoffman's chagrin, some geologists continue to use Slushball as a blanket term for all not-quite Snowball Earth scenarios—which glosses over the nuances that make some more viable than others.

Whatever you call it, the not-quite Snowball was at odds with the complete Snowball, and only more time with the strata would help settle this debate.

Plenty of other questions remained unanswered too, including what exactly had triggered the glaciations, what had ended them, and what had been their effect on the evolution of complex life then and thereafter. Lingering questions also remained about why we don't see any evidence of global glaciations in the rock record after the Cryogenian. Was the planet somehow predisposed to freeze over back then in ways it has not been since?

Theories were brewing, but those theories needed more on-the-ground proof, and that's what we were in Australia to collect. We would smack off chunks of rock samples to bring back to the lab, and fill our field notebooks with stratigraphic logs to try to make sense of the ice's dynamics.

During my first month in the Outback, I worked mostly one-on-one with a colleague of Adam's from Arkansas named Ryan Ewing, who years earlier had spent a summer working as a field assistant on Snowball Earth rocks in Namibia with Paul Hoffman, and who had also overlapped with Woody Fischer as an undergraduate at Colorado College. The world of Earth history, I was learning, is a small one.

By the time I met Ryan, he had come to specialize in studying how winds shape sedimentary landscapes as they sweep sand into dunes and other such features. This niche area of study offered a helpful angle on Snowball Earth because the Cryogenian would have been a very dry and windy time. The scattered pockets of bare ground would have, presumably, gotten whipped up into dune fields similar to those that rise and fall across the Dry Valleys of Antarctica today.

When mapped on a regional scale, the shapes and structures of dune fields can reveal something about the broader atmospheric patterns that formed them. Wind speed and direction orchestrate global air circulation, which, in turn, dictates the amount of heat that travels from the equator to the poles. Studying dune fields from the Cryogenian could help shine a light on the intricacies of the global climate that existed at that time. The Flinders Ranges contain some

ICE

of the world's most complete Cryogenian dune deposits, and thus offered the perfect venue for Ryan to conduct his work.

We planned to pop our tents on ranchland with permission from the local sheepherders who owned it, and follow our paper geologic maps and handheld GPS device to the outcrops of interest. This was shortly before the rise of smartphones, but even if we had had Google Maps at our fingertips, the patchy cell service of the Outback wouldn't have been strong enough to reliably guide us through the bush.

I had spent the prior summer floating in glacial waters collecting dozens of sediment cores to try to piece together a scattershot image of what that fjord floor looked like beneath us. Walking through the Flinders was like parting those glacial seas and walking out onto the very same types of sediments I had been trying to get a handle on, albeit hundreds of millions of years older. There was magic in seeing a fjord floor exposed like that, in all its relief. Traversing its contours, I imagined how that dusty topography mirrored what might have rested beneath our boats in Svalbard.

Though working in front of actual glaciers had come with more immediate thrills than traipsing through the remnants of ancient ones, something about piecing together those Cryogenian puzzles appealed on a more grounded level. We were able to sit and breathe and think about what we were studying without massive chunks of it crumbling all around us. What had happened had already happened, and we simply needed to piece the story back together.

I learned with time, too, that the Outback had thrills of its own. While sitting around a fire on one of our first nights camping, Ryan asked if I knew the trick of spotting spider eyes with a flashlight beam. I looked at him over the crackling eucalyptus branches and told him I did not. "Here," he said, placing down his mug of whiskey. "Let me show you."

We both clicked on our headlamps and scanned our beams past the shrubby land that surrounded our tents. If you shine the light at just the right angle, he told me, the beady eyes catch it and glow.

As we swung our beams around, dozens of eyes shone back at us, bright as stars. My blood ran cold.

From then on, I hoarded my shoes inside my tent each night. It was either going to be venomous spiders in my boots if I left them outside, or bits of dried-up sheep patties by my face if I pulled them inside. I chose the patties. I fell asleep to the dull thumps of emu and kangaroo around our camp—dirt beneath my fingernails and the smell of eucalyptus in my hair, ancient sand dunes sliding past my eyelids.

✦ ✦ ✦ ✦

UNLIKE BRIAN HARLAND IN his early days on Svalbard, we arrived in Australia with at least some inkling of how these ice ages may have begun and ended.

The onset of these deep freezes seemed, like the rise of oxygen, to be tied up with the continents. In this case, it was the distribution of the continents across the Earth that may have helped grease the wheels for change.

Geologists determine the ancient distribution of landmasses by studying the same magnetic mineral grains that Brian Harland used to figure out that his Svalbard rocks formed near the equator. These iron-rich grains swivel and align with Earth's magnetic field as they cool within magma, orienting themselves like fossilized compass needles. If magma forms near the equator, its magnetic minerals cool horizontally; if it forms near a pole, its minerals point vertically. If the magma lands somewhere in between, the minerals will align diagonally at an angle dictated by their latitude. Geologists access and measure the orientation of these minerals using a device called a magnetometer.

Harland and others during and after his time gathered many magnetic mineral samples from Snowball Earth–aged rocks to triangulate their latitudinal origins, and were surprised by what their magnetometers revealed: Over and over, they found grains that

ICE   105

rested relatively flat, pointing to origins near the equator. Together, the grains painted a pixelated picture of a planet whose landmasses sat huddled around its waistline, leaving its top and bottom sloshing around as empty expanses of open ocean.

This unusual clustering of land around the equator could have helped tip the planet into a global ice age for a few reasons, including a rather simple one: Open ocean freezes over more easily than open land does. That is, it's harder to produce enough precipitation to cover an entire continent in a thick layer of ice and snow than it is to freeze over water, even if that water contains salt.

With the top and bottom of the planet land free, ice could have easily and swiftly swept across these coldest reaches of the earth, brightening the surface of the globe and reflecting heat from the sun back out to space. As ice creaked farther from the poles, more of the ocean would grow brighter and, in turn, deflect more heat from the sun. Global temperatures would drop. Ice would beget more ice. A runaway feedback loop would ensue, not unlike the White Earth Disaster that Budyko's model had predicted. Frost and snow would eventually race all the way to the equator, easily enveloping the low-latitude land masses that, by then, had grown cold enough to accommodate such precipitation.

This premise for the onset of the Cryogenian has always charmed me with its simplicity. I find it almost comical to think that the conditions that helped precipitate the most extreme ice ages in all of Earth history may have boiled down to the shapes and colors present on the planet's surface at that time. It feels like a lesson that we could have learned in kindergarten, in the same way that young minds can easily grasp the concept of plate tectonics by noticing how all the continents fit together like pieces of a puzzle.

But, I have learned over time, nothing in the Earth system is quite this simple. There's always more complexity to be found under the hood. The arrangement of the continents around the equator may have established the conditions necessary to support a global freeze, but that arrangement itself wasn't the instigator of change. It couldn't have been if the continents had been arranged that way for millions of years before the Sturtian snowball got rolling, which a number of studies suggest was the case. Something else, then, must have nudged the planet's thermostat into disrepair, and what this was remains an active area of research and debate.

The path out of the deep freeze proves even harder to conjure from the rock record than the path in, but one leading theory offers another relatively straightforward explanation. Joe Kirschvink, the Caltech researcher who first coined the term Snowball Earth, recognized that the planet's tectonic plates must have continued to push and shove and pull away from each other throughout these ice ages. Volcanoes poked their peaks up above the ice and periodically puffed out warming gases. With most of the planet's landmasses and waterways covered, the sinks for these gases would have been largely sealed off. Like oxygen, carbon dioxide accumulated unabated. Eventually, over tens of millions of years, enough of this greenhouse gas built up to tip the planet back into a warmer state, possibly reaching as high as 660 times modern concentrations. This massive accumulation of carbon dioxide boomeranged global temperatures from far below freezing to sweltering, possibly above a global average of 120 degrees Fahrenheit, with sea surface temperatures reaching beyond 100 degrees Fahrenheit at the poles. Once the planet hit a certain threshold of heat, the snowball rapidly began to drip away over the course of thousands of years.

The melting that ensued cascaded in a feedback loop running in reverse of how the ice ages had begun. The disappearing ice revealed seas that hadn't seen the sun for thousands of millennia. The newly exposed dark open ocean begot more dark open ocean, absorbing more heat from the sun until all of the ice vanished. The chemical shock of the event primed the oceans to deposit the thick carbonate layers that often rest directly atop Cryogenian glacial deposits. Such "cap carbonates" are unique to Snowball Earth events.

Eventually, carbon sinks drained away some of the excess

greenhouse gases in the atmosphere and the planet's climate found a new stability.

At least, that's how the theory goes.

"Even though it is not something we would ever see or experience today, [Snowball Earth] is a great way of communicating how the Earth system interacts," says Jessica Tierney, a paleoclimatologist at the University of Arizona who was a lead author of the Intergovernmental Panel on Climate Change's Sixth Assessment Report in 2021. She doesn't study Snowball Earth herself, but is a strong proponent of efforts to draw on knowledge of the deep past to improve our understanding of what's to come. There is no doubt in her mind that the Cryogenian marked an extreme chill-down of the Earth system, she told me. That much is settled.

"What's more interesting is how it happened in the first place and how we get out of it," she said.

That's what we were trying to help work out in Australia.

✦ ✦ ✦ ✦

AFTER OUR FIRST WEEK in the field, Ryan and I made our way to a copper mine just outside the small town of Pernatty, not far from the shores of a desert playa called Pernatty Lagoon. Back in the 1980s and '90s, a few Australian geologists had identified the rocks in that mine as windblown dune deposits from the time of Snowball Earth, but they hadn't done much to study the conditions that those rocks formed within. They did, however, identify features that seemed to support the possibility of a soft Snowball, or Slushball, rather than a hard Snowball. Ryan hoped to shore up some of the missing pieces of the story that lay exposed there.

We arrived unannounced at a collection of trailers that appeared to be the mine's headquarters. A woman with cropped blonde hair greeted Ryan in front of one of the trailers and offered a few encouraging nods and a squinty smile as he told her who we were and what we hoped to do. She led him inside for a few minutes while I waited skeptically in the passenger seat of the truck, not convinced they'd grant us access to their rocks.

But when Ryan reemerged a few minutes later, he approached the truck with a pleased smile. "We're good to go," he told me as he settled back into the driver's seat. Not only had Christine, the mine's bookkeeper, given us permission to work there that week, but she had also offered us a small trailer to sleep in for the next few nights, a warm shower, and an invitation to join the crew for supper in the evenings.

We approached the rim of a large open pit with Willie Nelson's "On the Road Again" playing from the truck's speakers like an anthem. The part of the mine we would be studying wasn't in operation at the time, so we would have it mostly to ourselves, save for the bubblegum-pink galah cockatoos that screeched against the cloudless sky. Ryan parked the truck and we walked the rest of the way to the site down a steep dirt road into the base of the pit, where the rock rose up around us like amphitheater walls.

The layer-cake slice of red earth revealed troughs and humps of dunes that measured more than thirty feet thick in some places. Dunes like these emerge from seas of sand that, like seas of water, develop slopes shaped by the speed and direction of wind blowing over them. A strong enough gust will push grains up over a dune crest and down the leeward side, avalanching heaps of sand along with them. Over time, grain by grain and avalanche by avalanche, the entire dune will migrate in the direction of the wind. When preserved in the rock record, that motion appears as swooping concentric layers, which is what we found in the mine. If it had not been for the copper deposits discovered there, these layers would never have been exposed and made available to help piece together the story of Snowball Earth. Like the Soudan Iron Mine, it's an example of the paradox that understanding the history of the Earth has, historically, come at the expense of the Earth itself.

We pulled up the hand lenses that dangled around our necks to

I C E

take a closer look at the frosted gray and ochre sand grains embedded in the walls. They had been whittled into near-perfect spheres, far rounder than the angular sediments we had found within river deposits earlier in the week. The action of sand popping and hitting and bumping and rolling against other grains of sand in gusts of wind shapes them more uniformly than river water does.

After several minutes of close-up examination, Ryan nestled in his camp chair to draw sketches of the swooping cross section of the dunes from the middle of the mine. I went off on my own to walk along the edge of the walls, becoming dizzy from the fumes of the copper-leaching acids that had pooled to form a small pond at the mine's base. As I carefully stepped along that acrid shoreline and clung to the wall for balance, I began to notice a strange series of shapes in the strata beneath my fingertips. Whereas the layers on the far side of the pit that Ryan was sketching lay organized and parallel to one another, these strata were marbleized, with white and beige and ochre streaks all swirled together into what looked like a mushy triangle standing on edge with a broad top that narrowed down to a fine tip.

I continued along the edge of the pond, growing dizzier and more headachy from the fumes, and found several more of these swirly, downward-facing marbleized triangles, each as tall or taller than me. I rested my pencil down next to one for scale, snapped a picture on my digital camera, and walked back to show Ryan what I had found.

"Ah-ha," he said. "The wedges."

These were the features that had previously been invoked as potential evidence of a Slushball Earth rather than a Snowball Earth. Sand wedges form in modern polar environments during cycles of freeze and thaw. The ground buckles and cracks in a freeze, collects windswept grains within those cracks, and then heaves and swirls those grains as it warms and softens.

If the same process had formed the wedges in the mine, the ground they grew within would have needed to freeze and then thaw. But thawing ground was not an option for the fully frozen version of Snowball Earth as it had originally been conceived. Nailing down the timing of the wedges in the mine would prove critical to getting the story straight. If the wedges had formed at the beginning or the end of a glaciation, then they wouldn't pose a real threat to the hard Snowball narrative at all; freeze and thaw are expected of any transition into and out of an ice age. But if the rocks had formed directly in the *middle* of the glaciation, as the timing in the earlier papers had suggested, then the wedges would crumble parts of the hard Snowball narrative.

We filled our cameras with pictures of the wedges, drew out detailed sketches, and collected measurements that Ryan would analyze back in the lab to get a better sense of the conditions that those wedges formed within. While there was no datable material from within the mine to get exact ages of the wedges, material that others in the research group were gathering elsewhere in the Outback could help triangulate their timing within the broader Snowball Earth story.

As with most stratigraphic fieldwork, we didn't enjoy much glitz or glam of discovery in the moment. Only through follow-up analyses could we pull meaning from the rocks, and even then we would only be adding a few sentences to the story that others had spent lifetimes trying to work out. We each stood on the shoulders of those who came before us.

✦ ✦ ✦ ✦

AFTER SHARING A DINNER of lamb stew with the crew that night, we wandered with them to a nearby campfire and passed our handle of whiskey over flames that licked green and pink with chemical-soaked kindling. A steady flow of jokes and stories streamed along until the oldest of the miners stood up and faced his back solemnly to the group. Everyone grew silent. After a pregnant pause, the man creaked his head of long gray hair back toward us, cracked a toothy grin, and dropped his jeans to moon us all.

I C E

111

The rest erupted in laughter. "Come on," one of them said between chuckles. "Not in front of the lady."

I turned away from the mooner, burning with an awareness that I was not just the only woman in the circle but, aside from Christine, the only woman I had seen on the premises of the mine. I didn't really mind them having their fun, but I also understood that I would need to stand my ground. Ryan had been nothing but gracious and respectful during our time working together, but I had had my share of other disheartening experiences of sexism in science to remain cautious: the assumptions that I couldn't carry or operate field equipment given my small frame; the male-centric culture of brusqueness.

Even the men among other men faced their fair share of challenging dynamics in this field. Over time, I learned that Ryan hadn't received the same generosity of spirit that he offered me when he was in my shoes, working as a young assistant of Paul Hoffman on Snowball Earth rocks in Namibia. Ryan had been eager to work alongside such a hotshot in the field, but the experience turned out to be a rough one. When he misinterpreted the rocks, Paul would storm away in anger. When he was too quick to share an observation, Paul would tell him that he must not have gotten a very good education in college. Paul was notorious for this type of bullish behavior, not only with his field assistants but with students and colleagues alike.

It seems fitting, somehow, that the most intense climatic epoch in all of Earth history would appeal so deeply to a human with such an intense personality. In my experience speaking with Hoffman years later, that intensity seemed to well from a genuine love for and curiosity about the science, and an unflagging determination to get to the bottom of things.

"Snowball Earth is an example of the kinds of amazing things that Earth has been through that we would never have suspected if we didn't investigate the geologic record," he told me in his gravelly voice.

112 STRATA

He routinely interrupted me and spoke over me, but when he did so, it was always to further complete a point that bubbled out of him almost compulsively. He couldn't seem to contain his thoughts on how the planet worked. The endlessly fascinating inner workings of Earth seemed to come first for him. People, perhaps, came second.

"The history of our planet," he told me, "is one of the greatest stories."

✦ ✦ ✦ ✦

AFTER OUR FIRST MONTH in the field, Ryan and I met up with Adam Maloof, a geochronologist named Blair Schoene, and four students of theirs. Our days together fell into a comfortable rhythm. We woke to birdsong at sunrise, made coffee and a simple breakfast, and then gathered in a circle for a quick round of hacky sack before filling our packs with sample bags and water bottles and going off to spend the rest of the day out with the rocks.

I wrote wistful journal entries during this time, proclaiming that I could live that way forever. That there was nothing better than waking up to the melodies of desert birds and spending each day wrapped up in those ancient dramas, existing together as a sort of nerdy academic family.

A few discomforts did punctuate our days. The plump flies that landed on our cheeks and eyelids when the breeze died down. The thorny shrubs that released tiny burs into our socks and jabbed our ankles with every step we took, which we nicknamed "fucklets." But we adapted. We wore bandanas around our faces to keep the flies away and tucked our pants into our socks to keep the fucklets at bay. These were small prices to pay to bury ourselves in deep time and temporarily detach from the heartaches of the modern world. Never before in my adult life had I gone that long without reading the news or checking my cellphone or email, and never have I since.

I rode this high for a while, and I expected it to carry me through the entire field season. But about a month and a half in, as we bumped

ICE

along another dirt road to another outcrop, Adam powered on the satellite phone that we checked every few days for emergency messages from back home.

Music streamed from the speakers of the truck as I dug around the backseat for a bar of chocolate. When I turned to face forward again, Adam was looking at me with concerned eyes and reached back to hand me the phone. "You should call your mom," he said.

I held the cool, clunky thing to my ear as my mother's voice told me that my grandmother, her mother, had passed away a few days earlier. That the funeral would be in Houston within the month, and that I shouldn't worry about trying to make it back in time.

My body turned to lead as I began to suggest otherwise, but she wouldn't let me. "Just stay," she softly insisted.

My grandmother had been suffering from Alzheimer's for years by then. Her death didn't come as a particular surprise. But I still ached with the loss and the thought of missing this time with my family, with my own mother.

I powered off the phone and told the men in the truck what had happened. I told them I was fine, and I convinced myself of that too. I wasn't the only one missing out on time with their family. Blair had an infant back home. I understood this as the reality of conducting work in far-flung reaches of the earth.

But over the course of that second month, my stifled grief for my grandmother seeped into my daily routine and things began to wear on me that hadn't earlier in the season. The glow of the past was beginning to lose some of its luster, as was the adventure of the very long field season. I felt marooned 635 million years ago, looking for answers to esoteric questions that had seemingly so little bearing on our actual lives today. The meals we cooked together became less and less appetizing as the coolers we stored our food in became increasingly goopy. Not showering for weeks at a time had felt like some sort of badge of honor for the first month, but I began to crave cleanliness in a way I hadn't before, promising myself I'd never take another shower for granted.

"A clean pair of socks will change a whole day," I declared at the end of one journal entry.

During my final month in the field, I worked one-on-one with Adam's graduate student Catherine Rose, the only other woman in the group. I don't know if it was her quick British wit, the ease with which she cursed the outcrops, or the unspoken camaraderie of working with another woman for the first time in months, but I finally felt safe to share my grumblings. They came out in torrents, and she received them graciously. I sat and cried for a long time one evening, my salty tears mixing with the mediocre stir-fry we had cooked up from the dregs of our dirty coolers. I told her stories about my grandmother. She listened and I kept talking, kept crying.

I was twenty-two years old and grappling with the beauty and futility of everything. Of the moments that we live during our brief blips on this planet, and the moment that we die and become part of the strata we were studying.

✦ ✦ ✦ ✦

A FEW YEARS AFTER we returned, Ryan published his wedge findings with a handful of other colleagues, including Woody Fischer. He listed me as an author as well, though I did not do much beyond the fieldwork. According to measurements and calculations of the ways that modern wedges form, the Australian ones seemed to have developed during a time when average surface temperatures hovered right around freezing. This was more indicative of a soft Snowball than a hard one. But without datable rock material from the layers around the wedges themselves, it still wasn't clear whether they represented a time before, during, or after the glaciation.

It wasn't a groundbreaking paper, but few studies in Earth history are. It was a sentence added to the story that dozens of other researchers are still working to revise and eventually get straight, to the extent that the rocks available allow them to. "Nothing hurries geology," as Mark Twain once wrote. He wasn't wrong.

ICE

Proponents of a hard Snowball Earth agree that it may have had some sizable cracks, brought on by the rise and fall of Earth's crust as tectonic plates creaked beneath the ice. So long as about 99 percent of the ocean was covered in ice, the theory stood strong. The 1 percent left over would leave plenty of space to produce some of the confusing features found in the rock record, along with the oases that certain lifeforms may have needed to persist through it all.

While some researchers still find the whole idea of ice covering the planet hard to stomach, many have become more willing to entertain it as not only a possibility but a probability. Still, the research is not without its roadblocks. When I reconnected with Adam Maloof to catch up on where his work on Snowball Earth stood more than a decade after our time together in Australia, he was a bit downtrodden.

"It's a little stagnant right now," he told me. Two research papers that he had come across in the intervening years had thrown his understanding of his own research on its head. The papers examined processes taking place within modern carbonate platforms in the tropics, and indicated that he had, for years, misinterpreted sedimentological and geochemical data from his ancient carbonate samples. "Both of them just destroyed me," he said.

Very often, Earth science departments are split between those who study the past and those who study the present. In theory there is cross talk between these two areas, but in reality the walls of those silos often remain very thick.

Aware of the shortcomings of his interpretations, Adam has since shifted gears to make up for lost time and acquaint himself with the sedimentology and geochemistry of modern carbonate platforms in the Bahamas. He thinks he has strengthened his arsenal of tools enough to return to Australia with clearer purpose.

"I don't think they are not useful," he said of the samples we collected in the Flinders, "they just don't mean what we thought they meant."

There is still no question in his mind that the theory of Snowball

Earth is largely correct; it just needs fine-tuning. It's worth getting straight, he says, not only to understand this fascinating time in Earth history but, more practically, to help inform the climate modelers working to predict the future of Earth's rapidly changing modern environment. Such models have hundreds of knobs that work to re-create the planet's climate by breaking it up into its many different components—the flow of gases into and out of life and land, the collection and dispersal of heat in the oceans, wind speed and direction, solar luminosity. The list goes on.

While these models, as they stand, effectively simulate realistic patterns and shifts in the short term, Adam worries that they don't take into account some of the longer-term feedback loops that made the Cryogenian possible, and that are still at play within the Earth system. "There are memories in the ocean of times before the Industrial Revolution, which are still emerging into our climate system," he told me. Water warmed at the sea surface during medieval times has been ribboning through oceanic currents that are just resurfacing today. "Those memories are often neglected in climate models," he said.

This is what keeps him motivated to continue to improve our understanding of how the planet spun into and out of the Cryogenian.

Questions also still linger around the ice's role in the evolution of life then and thereafter. Some of the best traces of this post-Cryogenian shift in biologic complexity sit covered in sea spray on the southeastern coast of Newfoundland, along with remnants of the Gaskiers glaciation—the not-quite global ice age that followed 55 million years after the Marinoan, and that directly preceded the explosion of complex life.

I traveled there to witness the living, breathing world that Snowball Earth left in its wake.

Seven

THE LANDSCAPE OF COASTAL NEWFOUNDLAND HAS nothing in common with the Australian Outback. Replace the scent of eucalyptus with pine. Replace the dry, sharp breeze with soft moisture everywhere. Replace kangaroos and wallabies with moose and caribou.

Spruce and fir grow in thick, impenetrable clumps called tuckamores, all snaggled and stunted by hurricane-force winds that whip through in the winter. Moisture falls from the sky on more days than not, flooding bogs of heath moss with hidden puddles that soak the socks of anyone willing to traverse them. Off the coast, the Gulf Stream's warm breath barrels up from the south and collides with the much colder Labrador Current, knitting together a thick and persistent fog that drapes both land and sea. "This place . . . ," writes Annie Proulx in *The Shipping News*, "this rock, six thousand miles of coast blind-wrapped in fog." An estimated 10,000 shipwrecks have crashed along these shores since the sixteenth century.

For such a foreboding landscape and seascape, the place names come as a bit of a surprise. Along one stretch, an hour's drive will take you from Heart's Content to Heart's Desire, on to Heart's Delight and down through Dildo and South Dildo. Continue around to the other side of the bay and you'll pass through Come By Chance, Goobies, and Little Heart's Ease.

I traveled to these storied outskirts of Canada to meet up with Paul Myrow, a seasoned stratigrapher whose name I had heard thrown around quite a bit during my time in Australia. He was a beloved professor of both Ryan Ewing and Blair

Schoene, along with Woody Fischer—all of whom overlapped as students at Colorado College where Myrow has taught since the late 1980s. He was a common thread among these geologists whose work I had come to know well, and I had been intending to reach out to learn more about his own work. But before I could, I ran into him by chance at the St. Anthony Falls Laboratory in Minneapolis. He was in town collaborating with colleagues for the week and I was there on a reporting trip. I knew his face from headshots I had seen online, so when I saw him squinting over a laptop next to a series of whirring water pumps and flumes, I approached him.

"Paul Myrow?" I asked.

He lifted his head of curly brown hair with a smile, and I introduced myself as an old colleague of Ryan and Blair's. The world of Earth history is small enough that my name was familiar to him as well, and we quickly got to chatting. I told him that I was trying to find a way to get to the rocks in Newfoundland that were deposited soon after Snowball Earth, and I asked if he knew of anyone who might be able to help me. "Oh," he said with a spark in his eyes. He had spent years studying those rocks in the 1980s as a graduate student, he told me, and had spearheaded some of the research around them that I was interested in learning more about, including the Gaskiers deposits and the following explosion of complex fossils that appears in the Cambrian.

It felt like kismet. Within minutes we were making plans to meet up on those very rocks the following year, when he was slated to co-lead a field trip with students and faculty from Syracuse University. With a quick email to the organizers of the trip, he arranged for me to tag along, and the next summer I boarded a plane to meet them all there.

Our itinerary was more geologic sightseeing than field research. It was a chance to immerse ourselves in the post-Snowball world and commingle with the beings that emerged as a result. To see and touch their physical remnants and, through this connection, begin to grapple with the conditions that made their lives—and, in turn, our own lives—possible. How had Snowball Earth paved the way for the emergence of multicellular life? And how did these first widespread multicellular beings survive this newly thawed world?

More than half a century after Brian Harland first began pondering these questions, some answers were beginning to coalesce. But the future of the research relied on the minds and dedication of younger generations of scientists. The participants of the field trip were mostly undergraduate and graduate students in Syracuse University's Department of Earth and Environmental Sciences. Guided by the more seasoned researchers leading the trip, we would walk strata that trailed out of the Cryogenian and into the Cambrian. We would bear witness to the explosion of life, get up to speed on what we know about these rocks today, and learn what work still lies ahead for future generations to tackle.

Where Paul Hoffman might come off as intimidating and cerebral, Paul Myrow is as disarming and sociable as they come. At sixty-five years old at the time of our trip, he had five folk-rock albums under his belt and the energy of someone decades younger, though he had never owned a smartphone.

The trip's lead organizer, Linda Ivany, was a similarly charismatic marine paleoecologist from Syracuse University who had studied with the famed paleontologist Stephen Jay Gould as a graduate student at Harvard in the 1990s and spent her spare time riding horses. Where Paul Myrow's focus was the strata, Linda's was the fossils within those strata. Together, they made a stratigraphical dream team.

We spent our first morning on the pine-covered shores of Conception Bay, looking for the unsorted mix of sands and cobbles that made up the tillites of the Gaskiers glaciation. Like the full-fledged Snowball Earth glaciations of the Cryogenian, the Gaskiers appears to have been widespread across at least eight paleocontinents and has a carbonate layer associated with it in some places. Unlike the

120   **STRATA**

Sturtian and Marinoan glaciations, however, the Gaskiers lasted only a few hundred thousand years. That is not quite long enough to spin into and out of a full-fledged global glaciation. Instead, this glaciation may have begun creeping toward the equator but then crept back before it could overtake the entire planet.

The conclusion of the Gaskiers marked the end of an era in more than one way. Never again do we see evidence of a global or near-global ice age in the rock record, and never again are single-celled beings the lead characters on Earth.

As we pulled down a quiet, tree-lined residential road to look at Gaskiers deposits that Myrow had studied decades earlier, a local gave us a friendly wave and the go-ahead to park alongside his driveway, seemingly unfazed by our interest in the rocks around his home. We poked through a few other people's lawns looking for the right outcrop, until we finally found that telltale glacial mix of beige and gray cobbles and pebbles and sands. As we stood in a circle, Paul brought the students up to speed on the story of Snowball Earth, and then he sent us wandering around to look for the smooth, cream-colored carbonate layer that capped off the glacial deposits. After poking around half a dozen backyards trying to find the end of the Gaskiers, we realized that we were struggling to find it because it was submerged beneath the high tide.

Having seen all that we could of the last layers of this almost–Snowball Earth, we caravanned on to deposits laid down in the aftermath of the ice.

✦ ✦ ✦ ✦

OUR NEXT DESTINATION PROTRUDES like a giant barnacle off the southern coast of the Avalon Peninsula, poking a bit farther out to sea than the surrounding coastline. Before the advent of modern navigation, mariners often mistook this point for another one around the bend, earning this place the name Mistaken Point. The numerous shipwrecks that resulted from mistakes like these account

I C E   121

for this region's more morose nickname, "the graveyard of the Atlantic," which applies not only here but for much of the southern half of the Avalon Peninsula.

On the pale gray rocks of Mistaken Point rests another graveyard of a different kind—the fossilized remains of the planet's first widespread group of complex organisms large enough to see with the naked eye. The evolutionary dice that rolled and spat them out produced all sorts of odd shapes and forms only the likes of Dr. Seuss could think up. Some grew long and feathery, others sat on narrow stalks, some splayed out as oblong ovals, and still others lay like chewed-up fish carcasses. None had mouths or anuses or bones or shells. They existed together as a community of soft-bodied globs on or near the seafloor.

With their unprecedented size and shapes, these beings opened up a whole new playing field of biological complexity that the planet had not yet known. I say "beings" because paleontologists often don't know what bucket of life to throw them in. Were they algae? Animals? Fungi? Bacteria? Some failed biological experiment that concluded in an evolutionary dead end? More often than not, they more closely resemble each other than any living group of organisms today—though some do appear related to sponges and cnidarians such as corals and jellyfish.

Upward of 200 different types of these soft-bodied enigmas crop up around the world, from the White Sea region of Russia to parts of Namibia, Iran, Australia, and elsewhere. They fall within rocks that date between about 579 million to 538.8 million years old, during a geologic period named the Ediacaran, for the Ediacara Hills of Australia where they were first described in detail in the 1940s. This period paves the way for the larger explosion of even more complex life and true animals that come with the Cambrian, the next period on the geologic timeline.

Of all the locations around the world with Ediacaran deposits, Mistaken Point contains the oldest and most extensive collection of fossils, and for that reason has been designated a UNESCO

122

STRATA

World Heritage Site to protect them. To visit, you must arrange for a guided tour with staff at the Mistaken Point Ecological Reserve, a seventeen-kilometer stretch of coastline with more than one hundred fossil surfaces, a motley grab bag of passports back to deep time.

We started our tour with the oldest strata in the park, 579-million-year-old layers of seafloor that held the very earliest traces of these post-Gaskiers beings in the world. Our guide, Mark, led all nineteen of us down a narrow stretch of beach to a slab of light-gray bedrock that contained a series of round impressions with bubbly textures. As I traced my fingers along one, I expected to feel moved by these earliest of Ediacaran biota. But, if I'm being honest, I wasn't quite sure what to make of the thing. It didn't really look biotic to me. Its form did not have any intricate indentations or patterns that I associate with life, but instead sat amorphous, like a fossilized puddle of gurgling mud.

Paleontologists studying these forms apparently agreed that they didn't quite look alive; they, more creatively, named them "the pizza discs." A layer of white, silty volcanic ash sat above the exposed surface, and it's that ash bed that preserved the imprints on the seafloor, much in the same way that Mount Vesuvius preserved the ancient city of Pompeii. The ashes also gave a whitish hue to the pizza discs, a topping of "mozzarella" if you will, further justifying the metaphor.

Without the ash, the traces of the pizza discs would have become muddled and erased from the fossil record before the seafloor turned to stone. Ash beds like this exist across the park, and they are responsible for preserving all the soft-bodied imprints there. Volcanic ash also, fortuitously, often contains loads of the mineral zircon, which can be used to measure the age of the rock through radiometric dating. Because the ashes smothered and killed the biota beneath them in a flash, the ages of the zircons collected there should align exactly with the ages of the fossils themselves.

As I continued poking around at the discs, I couldn't help but try

ICE

123

to put the forms into some sort of categorical bucket of life, even though I knew it was a lost cause. Everyone else in the group seemed to feel the same way. We all voiced our best guesses.

"Bacterial mats?"

"Some kind of fungus?"

"Dead clumps of algae?"

"None of the above?"

We were met with shrugged shoulders and knowing smiles from the more experienced among us.

"That's the question we will be asking the whole time we're here," Linda said to the group as we gathered around in a circle. "What are these things?" And what might they have to do with the end of the Cryogenian?

Some researchers have suggested that the pizza discs don't represent the same organism in every instance but, instead, are the mucked up, decomposed remains of a number of different organisms all too deformed to tell apart. Whatever they might be, Linda told us, they appear surprisingly shortly after the Gaskiers glaciation.

"Not just shortly after," Paul chimed in, "a *million years* after."

A blink of a geologic eye. For nearly 4 billion years, only very simple and very small beings knew Earth. Once the last of the Gaskiers ice melted, the pizza discs began cooking.

We continued up the coast to a nearby trailhead to visit with more Ediacaran fossils that formed some millions of years later, when evolution had rolled more dice and generated more intricacy. On the way down the path through heath and moss, we snacked on cloudberries and blueberries that grew among sprigs of cinnamon fern— the fruits of the labor of the half-billion years of evolution that had unfolded since the end of the Gaskiers.

About a mile in, we followed Mark over a trickling stream and down a hill onto a rocky shoreline pockmarked with tide pools. At first I couldn't see anything of note in the rock he pointed to. But as I squinted and ducked around to view the surface from different angles, I began to make out some shapes and forms. I pressed my

124   STRATA

hand down on what looked like an impression of a fish carcass, if the ribs were made of fern leaves and the fern leaves ended in feathers. I stepped back and noticed that they were everywhere, scattered around in lengths ranging from the size of my thumb to the size of my forearm.

A seal bobbed in the waves behind us, watching wide-eyed as the students clambered around with hand lenses and cameras.

For centuries, this bit of coastline had been home to seasonal and year-round communities of cod fishermen who traveled here from Europe and returned back across the Atlantic to sell their catch. The stream that we had crossed was the last source of fresh drinking water available to them before their transatlantic journey home. The children of those fishermen played among those stones and, the lore goes, called the impressions "flowers in the rocks." They probably wove together all sorts of tales about how those blooms got stuck there, while their parents splayed out their fish to dry on the hill above, too consumed with survival to give the rocks much thought. It wasn't until the late 1960s when students from Memorial University of Newfoundland recognized the significance of the shapes that they became more widely known and appreciated as some of the very earliest traces of multicellular life on Earth.

This surface was merely a teaser of what lay ahead. We climbed back up the hill and walked for a few more minutes until we reached a collection of wooden posts draped with rope where Mark told us to remove our shoes.

We lined up our boots and sneakers mostly in silence, as if preparing to enter a sacred space. The park only permits socked feet beyond that point to minimize scuffing.

As I stepped onto that expanse of ancient seafloor, I felt myself transported to the inky depths of an Ediacaran ocean. Dozens of undeniably biological forms lay splayed all around us, ranging in size from a finger to an arm, many as pronounced as footprints in wet cement. The flood of magic that the pizza discs hadn't quite

ICE

125

summoned came rushing in. Waves splashed below us, metronomes beating onward as we tumbled further back in time.

The hundreds of millions of years that separated us from the Ediacaran beings dissolved as I traced my finger along a narrow stalk that opened into a frond that resembled seaweed swaying in a current. My mind wished to paint it green, render it as smooth as kelp. But I batted away those urges to categorize them as quickly as they arose. Palm pressed softly to the stalk's edge, I accepted the invitation to imagine a version of Earth far different from our own. A version without backbones, without hair. Without coffee or newspapers or politics. A version that, for some 40 million years, allowed this motley assortment of mouthless, boneless beings to reign over the kingdom of life.

We roamed around, periodically huddling in circles trying to make sense of what we saw. Some of the beings looked like heads of lettuce, others like feathered fish carcasses. Again and again, we swatted away urges to put them in familiar biologic buckets. We crouched on our hands and knees, trying to avoid sitting down and scuffing the fossils with the rivets on our back pockets.

Once we had made our rounds of the tennis court–sized slab of ancient seafloor, Linda gathered us in a circle to discuss the latest paleontological understanding of these forms. As with the pizza discs, a lot remained unknown. In many cases, it's not clear if or how they moved—whether they sat stationary on the seafloor, as corals do, or if they propelled themselves around, more like jellyfish or sea slugs. It's also not clear in what way they were multicellular. Paleontologists assume they were, since they were too big to be one single cell. But were they multicellular in the way that we are, with cells communicating with one another through complex networks? Or did they have clumps of simple cells that didn't communicate much at all?

Stephen Jay Gould described them as pancakes, ribbons, and threads—forms that would have been flat enough to absorb all that they needed from the sea around them by way of diffusion.

126   STRATA

"And that probably works great," Linda told us, "as long as there is nothing else out there that is three-dimensional that can come along and eat you up or burrow through you."

A flock of gulls exploded over us with high-pitched calls, reminding us of the webs of predation that have evolved since then.

"This is the most extraordinary group of fossils that people have not been able to come to grips with," Paul said as the afternoon sun drifted lower toward the horizon, the angled light bringing the fossils into even sharper relief.

"I hope you all realize how magic this is," Linda told us.

As the scientific community continues to try to make sense of Ediacaran fossils, they have begun fighting a battle against time. In the past several decades, the fossil surfaces have visibly eroded as the intensity of coastal storms has increased with climate change. While some new fossil surfaces have emerged with this erosion, others are forever lost to the sea. Researchers are now scrambling to capture replicas of the impressions so that they may continue to learn from them long after they are gone. Some are scanning the rocks with a type of remote sensing called LiDAR; others are creating molds. One group from the University of Cambridge has come out with cameras at night, shining lights at angles that the sun wouldn't naturally hit to pick up on otherwise imperceptible details.

While these reproductions might help with scientific research, they are no replacement for the feeling of standing shoeless on the Ediacaran seafloor, in the company of those pioneering lifeforms, these enigmatic innovators of a newly thawed world.

✦ ✦ ✦ ✦

AFTER SOME 40 MILLION YEARS, these globs begin to drift out of the fossil record. It's not a knife-sharp extinction so much as a steady gradient toward mouths and anuses and eating and burrowing and other such things that those simpler sacks of cells could never quite muster. With this transition out of the Ediacaran comes the

I C E   127

Cambrian explosion of life, the beginning of almost every animal group alive today.

"Since then," writes Stephen Jay Gould in *Wonderful Life*, "more than 500 million years of wonderful stories, triumphs and tragedies, but not a single new phylum, or basic anatomical design . . ."

Because of what it meant for the evolution of life then and thereafter, many Earth historians consider this to be the most important transition in the entire geologic timeline. The southern coast of Newfoundland's Avalon Peninsula is, according to the International Commission on Stratigraphy, the best spot in the world to witness this transition.

Paul Myrow played a role in bringing these rocks into the global spotlight. He told us how this came about as we all stood huddled against a wet wind outside the small fishing town of Fortune, where a lighthouse rises up out of a collection of sandstones and siltstones called Fortune Head. There, a 440-meter-long stretch of coastline straddles the end of the Ediacaran and the beginning of the Cambrian. By walking those rocks, you can travel through the aftermath of the Gaskiers glaciation and straight into the swell of the Cambrian.

The International Commission on Stratigraphy designates official points of reference for geologic transitions like this one to help geologists connect the dots elsewhere around the world. Global Stratotype Section and Points (GSSPs), also called golden spikes, are singular packages of strata that best represent the transition out of one geologic period and into another.

To be designated a golden spike, an outcrop must be adequately thick, contain visual evidence of major changes in the fossil or geologic record, and be accessible to the public, to name just a few of the criteria. Geologists submit proposals for locations they think deserve this coveted title and then subcommittees gather to vote on one. They are as much points of scientific correlation as objects of national pride, sometimes even turning into revenue-generating tourist attractions—as is the case for Fortune Head.

When Paul Myrow was a graduate student at Memorial University in the 1980s, relatively soon after GSSPs first became a formality, a golden spike had not yet been assigned to the Precambrian–Cambrian boundary. As this is arguably the most significant transition in the whole geologic timeline, it needed one. But where, exactly, should that line be drawn? And was there a good contender at Fortune Head?

This was before Paul Hoffman had begun spreading the gospel of Snowball Earth, so Myrow was not looking at the rocks from the perspective of their proximity to those ancient glaciations; that part of the narrative had yet to take shape. He was more interested in the rocks from the perspective of what they might reveal about the onset of the Cambrian.

Pondering the rise of complex animal life along the rocky shores of Newfoundland might sound like an idyllic way to spend a few years of graduate school. By many measures, it was. But this location, Paul told us, also had its downsides. Back in the 1980s, it was the site of the Fortune Dump. The town piled up towering heaps of trash each week, lit those heaps aflame, let them burn for a few days, and then bulldozed the charred messes over a cliffside and into the ocean to begin the cycle anew.

To get to his place of work, Paul hiked through clouds of burning sofas and dodged refrigerators falling off the cliffside. He wore a mask to protect himself from smoke. In the end, though, these hazardous working conditions proved worth tolerating. In a layer of gray sandstone in a small cove barely beyond the dump, he and colleagues Guy Narbonne from Queen's University and Ed Landing from the New York State Geological Survey found a feathered pattern that alternated from the left and right of a central rib. This was *Treptichnus pedum*, a fossilized burrow that was known, from places it cropped up elsewhere, to be Cambrian in age.

Together with colleagues at Memorial University, Myrow decided that this appearance of *Treptichnus pedum* was a decent marker of the beginning of the Cambrian, since up until that point it had never been found in rocks older than the Cambrian, and it sat just above rocks that were identifiably Precambrian. Myrow and his colleagues

ICE 129

put forth a proposal for the Fortune Dump (now called Fortune Head) to be designated the Precambrian-Cambrian golden spike, alongside two other contenders from China and Siberia. It was a lengthy and sometimes bitter selection process, with national honor at stake, Paul told us as wind rattled our wet raincoats. He recalled stories he heard about a meeting where scientists from one country took the slides of another contender from a different country and inserted them into a projector with force, intentionally jamming and breaking them.

After years of deliberation, Fortune Head won by a narrow margin. The onset of the Cambrian finally had its golden spike. The town scrambled to move the dump inland, level the ground, cover it with vegetation, and designate the area an ecological reserve. A visitor center was erected downtown that now draws hundreds of visitors annually.

We followed Paul across the grassy top of the former Fortune Dump, where mangled metal and melted plastic still fall into the cold sea. When we reached the next cove over, we formed a single-file line, crouched on our bums, and scooted down the steep slope out onto the wet gray rocks of the Ediacaran's end. With the Atlantic Ocean splashing to our right and the Cambrian rising up in the rocky cliff to our left, we stood at the boundary of the Precambrian and the Cambrian. The *Treptichnus pedum* that Myrow had championed as the golden spike had rested just above where we placed our hands along that cliff, but had since been hammered out for closer study. Other *Treptichnus* fossils had also since been found in rocks below this, which had for a time called into question where, exactly, the boundary between the Precambrian and Cambrian really sat. Just as every stroke of midnight is not inherent but human made, this marker of time was also a construct.

What wasn't a construct was the presence, above where we stood, of some of the earliest evidence of a diversifying biotic world, energized by conditions left behind by Cryogenian ice.

After taking a moment to revel in the transition beneath our feet, we scrambled back up to the top of the cliff and walked out to the younger Cambrian strata we had just been standing beneath. There, we found corkscrew-shaped burrows no larger than the tip of a marker—traces of hiding and self-preservation that had not been possible, or necessary, before then.

I placed my finger on one of the corkscrews and tried to imagine the creature who formed it, but my attention drifted to a whale making waves in the distance. We had noticed splashes when we first arrived there and had been delighted, only to discover that what had first seemed like a lively display was, instead, a struggle for life. Through binoculars, the students in the group discovered that a piece was missing from the whale's tail, perhaps from a bite or entanglement with fishing gear.

As I watched the animal thrash around, the warm thrill of time travel drained quickly into a cold grief. Whether or not the whale's struggle was human caused, their thrashing felt like an emblem of this moment, the loss of so many lifeforms that had been evolving since the Cambrian that were now drowning away due to the short-sightedness of so many humans. I wished we could ease the whale's distress but, knowing that we couldn't, my wish morphed into a sort of nostalgia for the Cambrian, when the slate of hearts to break was still clean and there was nowhere to go but up.

As I stood to look around for more corkscrews and distract myself from the whale, my longing for the Cambrian morphed into a nostalgia for the Ediacaran, when multicellularity was even simpler, confined to the dark, sunless currents of the ocean's bottom. When everything remained soundless, scentless, sightless, and there was nothing to do but float and sway.

From there, I slipped into a longing for the Cryogenian, when life was so unimaginably hard and yet living things found ways to soften and push forward and persist. How easy it would have been for those microbes and algae and protozoa to give up. And yet they didn't and, because they didn't, we get to walk the earth and read their stories.

As we shivered our way back to the vans at Fortune Head, the

I C E 131

traces of the early Cambrian sat unblinking beneath the rain, telling us with a wordless wisdom that there are beginnings and that there are ends and that the fibers of the planet will always harden and soften and dissolve and re-form anew. That our own legacy will, some day, erode back into the sea.

But we're not here to consider our own end, at least not quite yet. We're just getting started acquainting ourselves with the beginning of life as we know it. Life that grew onward and upward out of the sea and, for the very first time, made a living on land. Life that turned a world once covered in ice into one covered in earth.

PART III
MUD

Before mud, land lay bare as stone. Continents rose up like suits of armor, hard and inert against restless young tides.

Over time, heat and frost and snow and rain etched cracks into that armor and crumbled it down. Boulders broke to cobbles to pebbles to sand, silt, and clay. These finest of grains fell into channels that swept out to sea, leaving land a dustless terrain of emptiness, winds wailing over a rootless Earth.

Orbits spun on. Glaciers licked out and in; species rose and vanished. All the while, continents remained stony and largely lifeless.

Finally, new cells emerged with a different whim. They grew stems and leaves; they bent and waved. They reached for the sun, bore down through the crust and morphed it from rock to refuge.

Out came the mud, out came the soft wriggle of life across land.

Eight

LET ME BE CLEAR: MUD HAS BEEN HERE SINCE THE beginning. Long before there were global glaciations or cyanobacteria puffing oxygen into the atmosphere or any bacteria doing anything at all, there has been mud. Bedrock has eroded down to the silts and clays that constitute mud ever since there has been anything available to erode.

But for most of the planet's existence—that is, for more than 90 percent of it—these smallest of sediment grains remained largely absent from land, accumulating instead at the bottom of the sea. It wasn't until about 458 million years ago—more than 80 million years after the onset of the Cambrian explosion—that mud finally began to dress the continents.

If you reach out your arms and imagine Earth's 4.54-billion-year history as a timeline that extends from the tips of your right hand to the tips of your left, we have now reached the area just before the heel of your left palm. The rise of mud on land.

When geologists say "mud" they mean the very finest bits of Earth's dandruff that stick together when they get wet. Pieces of dead plants and other decaying material can also make up mud and, in high enough quantities, can turn that mud into soil. But, geologically speaking, mud simply refers to a collection of tiny sediment grains, whereas soil refers to a combination of those small sediments and life. All the rest of the sediment sizes—the progressively larger sands, pebbles, cobbles, and boulders—don't stick together when they get wet, at least not on their own, so they don't constitute mud.

The story of mud's rise on land begins with the rise of land plants. These earliest sprouts descended from green algae and anchored to the ground not with roots but with hairy protrusions called rhizoids. What began as spongy, diminutive growths some 458 million years ago—early relatives of mosses and hornworts and liverworts—steadily spiraled out into sturdier and larger things with branches and leaves. By about 390 million years ago, they formed what we'd recognize as forests.

As plants greened the continents, they transformed every facet of the Earth system in both visible and invisible ways. They added kindling to the landscape and a burst of oxygen to the air that, together, ignited the very first wildfires. That burst of oxygen also spurred the evolution of new animals that grew ever larger and more energetically complex. Fish that once grew no bigger than golf balls evolved into species that swelled larger than balloons. They became better swimmers and better predators and, as it turns out, better at having sex. Internal copulation arose at this time in a species named *Microbrachius dicki*, not in reference to the male appendage, but for Robert Dick, the nineteenth-century Scottish geologist who discovered the species. (The fish's sex organ was discovered much later by Australian paleontologist John Long and his team.)

As oceans grew friskier and flames licked across the continents, plants fundamentally changed the ground from which they sprouted. Growths that grew up along the banks of rivers, drawn by the promise of water, added a roughness to the land that finally provided silts and clays something to cling to. And cling they did—both to the vegetation and to other particles of silt and clay. All those bits of dandruff that had long drifted out to sea began, at last, to stick together in mucky deposits within those greening banks of rivers.

And so with the rise of plants on land came the rise of mud.

The newfound muckiness of riverbanks didn't only change the texture of the earth, it changed the shape of topography and the ways that rivers flowed across the continents. Before mud accumulated, river channels ran mostly sandy and loose and chaotic, like rivulets at the edge of a beach. Those wide, sandbar-strewn channels lacked cohesiveness, and so when floodwaters rose, their banks heaved and collapsed and formed anew to accommodate the surges of water.

The arrival of mud brought newfound order to this chaos. Channels that once ran wide and wild with multiple threads narrowed into more organized, single-threaded sinuous channels such as parts of the Amazon and Mississippi Rivers today. These sturdier channels stood up against floodwaters, forcing surges up over their banks and out onto the surrounding land.

As those floodwaters reached their tendrils out beyond the beds of those sinuous rivers, they draped the ground in nutrient-rich silts and clays. Flood after flood brought drape after drape of these fine sediments that eventually built some of the planet's first floodplains that provided the foundations of the very first forests.

And so with the rise of mud on land came the rise of plants. The two went hand in hand.

Mud's arrival didn't touch only the physicality of the planet, but also the contents of the atmosphere and, in turn, the inner workings of the climate. The clay constituent of mud contains an electric charge that attracts organic detritus and traps it underground. In so doing, clays prevent this carbon-rich material from decaying and releasing carbon dioxide into the atmosphere. Mud's rise on land formed a powerful new sink for this greenhouse gas and a new knob on the global thermostat. (This smallest of mineral classes plays an outsized role in our global climate today, trapping the equivalent of some 20 percent of human carbon dioxide emissions every year.)

As muddy, carbon-storing floodplains sprawled across the continents, they created the first environments where animals could eke out a living entirely on land, away from the confines of water bodies. Evolution ensued. Out came land-bound legs and claws and hair. Fast forward half a billion years or so and early human civilizations began coalescing on the floodplains that lay beside muddy rivers. They farmed these lands that were naturally fertilized by the nutrients contained within mud, and they cooked and stored food out of earthenware vessels made of mud. Some built homes of mud.

Today, our relationship with this material transcends these practicalities and reaches into the realms of religion, philosophy, poetry. We recognize this material, metonym for earth, as a source of wisdom, a reminder of the beauty hidden within the seemingly ugly or mundane. "I want you to fill your hands with mud, like a blessing," writes Mary Oliver in her poem "Rice." "Without mud, there can be no lotus," writes Thich Nhat Hanh in *No Mud, No Lotus*, his book on suffering.

We're not alone in our connection with this substance. Without it, burrowing tadpoles and turtles would have nowhere to wait out the depths of winter at the bottoms of frozen lakes and ponds. Phoebes, martins, and magpies would have nothing to plaster their nests with. More than 10,000 species of mud wasps would have no way to lay their broods, and elephants would have nothing to sling across their bodies on the hottest of days.

For a material as widespread and seemingly synonymous with earth as mud is today, it might come as a shock to learn that it was largely absent from the surface of the planet for more than 90 percent of its existence.

"Mud is one of the most common, abundant things you can think of," Woody Fischer told me of this late transition to a muddy world. "The recognition that for most of Earth history it was not like that is a big deal." A big deal for our capacity to conceptualize all the changes the planet has spun through in the past, and what changes might yet arise in the future. A big deal for our capacity to understand how the Earth works.

Neil Davies, a sedimentary geologist at the University of Cambridge, has led the bandwagon on the quest to map mud's rise through geologic time. By shining a light on the planet's smallest sedimentary particles, he has helped illuminate their outsized role in the Earth system in the past and present.

"Once you take it out of the equation and imagine the world

MUD

without as much mud on land," he told me in my first phone call with him, "then it becomes a very different kind of planet."

✦ ✦ ✦ ✦

NEIL DAVIES FOUND HIS way to mud a bit circuitously. After earning his doctorate in geology in the early 2000s, he set off to study a pile of fossilized fish that lay with mouths agape along a dusty roadside in central Bolivia. The fossils had formed sometime during the Middle Ordovician between about 470 and 460 million years ago, not long after the very first fish evolved—and, with them, the very first backbones.

Neil hoped that the sediments in and around the fish would help settle a long-standing debate about the environments that these earliest of vertebrates inhabited. Did this debate have anything to with mud on land? Not directly, but it's an example of how one line of scientific inquiry can, unexpectedly, lead to others that kaleidoscope together to broaden our understanding of the world we inhabit. Without our more obvious interest in the rise of vertebrates, we may not have gained any sense of how long continents remained mudless.

Most signs in the strata pointed to saltwater origins for fish, and the majority of the scientific community viewed this case as closed. But a minority argued for freshwater. Greg Graffin, the front man of the punk band Bad Religion, who earned a doctorate in zoology at Cornell University between gigs, had studied a collection of early fish fossils in Coloradan strata that he interpreted as fluvial—that is, laid down by a river or stream. He came to this conclusion based, in part, on the presence of coarse-grained conglomerate deposits similar to those you might find within the pebbly bed of a river today. He invoked these conglomerates as evidence that the earliest vertebrates swam through freshwater, not saltwater, and published a paper arguing as much in the *Journal of Vertebrate Paleontology* in 1992.

But others disagreed with his interpretation of the strata. Coastal seafloor sediments can share some of the same qualities as certain fluvial deposits, and distinguishing the two is not always straight-forward. Because all the other fossils of these early fish appeared in strata with finer-grained deposits that looked more definitively marine, the argument for freshwater origins in this one instance was a hard sell.

Graffin's theory was already unpopular when Neil arrived in Bolivia a decade later, but no one had managed to gather enough evidence to solidify the case for marine origins of fish. Neil and his colleagues hoped their findings might finally tip the scales.

After filling his field notebook with measurements and sketches of grain sizes and textures from that fossiliferous roadside, he and his team agreed that those fish had likely lived in a shallow coastal envi-ronment. They came to this conclusion based on the fine-grained, thinly layered nature of the rocks—qualities typical of many seafloor environments today. One more tally for marine rather than freshwa-ter origins of vertebrates.

Though those fish may have flapped through marine waters, the circumstances of their death appeared to have been tied up in freshwater flowing off the land. Their mouths and gills looked as if they had been violently stuffed with silts and sands that had washed offshore from a nearby river, perhaps during a storm. Similar gill-stuffed fish crop up in ancient coastal deposits in Australia, Oman, and elsewhere around the world in rocks of a similar age, suggesting this type of coastal suffocation may have been a somewhat common occurrence at the time.

Unlike rivers today, these ancient ones had no vegetation growing along their banks. The evolution of land plants was still millions of years away. Without roots clamping down on riverbanks, channels clouded with debris that they carried to their mouths and out to sea.

"You don't have this kind of stabilizing vegetation, so a lot of stuff is washing off," Davies told me.

Magnify this effect globally, and the results would have been profound—both for the surface of the continents and the creatures that swam beside them. Once plants greened the continents and river

MUD 141

channels firmed up, such fish suffocations probably became less frequent, Neil says—though they can still happen in some places today.

Those asphyxiated Bolivian fish got Neil interested in the nature of early rivers and how they evolved through time. He moved on to research how channels changed course with the rise of plants on land and, in turn, the rise of mud on land.

This interwoven plant-mud-river narrative had been apparent in the rock record long before Neil dug into this line of research. As early as the 1940s, a Norwegian geologist studying rocks on Svalbard first noticed that river deposits from Earth's earliest days tended to be sandier, less muddy, and less cohesive than those that formed after the continents greened. When plants arrived on land, the amount of silky, fine-grained mudrock notably increased in river deposits around the world.

This struck plenty of researchers as curious, but nobody did much with it for a long time. Decades passed, with the rise of mud remaining nothing more than a footnote to the seemingly more intriguing story of land plants. More glory could be found in publishing papers and delivering conference lectures on the oldest-ever vegetation than in investigating the effects that those sprouts had on the ground they grew from.

It took Neil Davies to finally bring the mud angle of the story into focus, in the past decade or so. Or rather, it took his sharp-witted and highly caffeinated graduate student, Will McMahon, to do the painstaking work of compiling more than 1,000 research papers and conducting more than a hundred original field investigations to measure the proportion of mudrock across all 704 known river deposits spanning from 3.5 billion to 300 million years ago—with Davies's dogged support and guidance along the way.

This tedious team effort revealed a major surprise: Across the 99 million years when plants evolved from mosses to trees, mudrock increased by more than tenfold. The proportion of these smooth, fine-grained deposits rose from as little as 14 percent in the oldest of their samples to as much as 90 percent in the youngest ones.

Neil and Will had expected to find some increase in this material through time, but not one as pronounced as this. They also hadn't expected the onslaught of mud to begin as early as the arrival of moss- and liverwort-like plants, before the stabilizing power of roots and stems came into play.

The finding raised some fundamental questions about the evolution of terrestrial environments. What accounted for this exponential rise of mud? And what role, exactly, did plants play?

Such questions may sound somewhat trivial. A bit granular, if you will. Does it really matter how these tiniest of grains began sticking around?

For Neil, Will, and others in their realm, the answer is a resounding yes. By nailing down the role that plants played in the early rise of mud, they hoped to unlock our understanding of the forces that have sculpted global landscapes and climate for the past 458 million years.

Before the rise of vegetation, mud formed mostly through physical erosion—by, say, the force of a rock tumbling down a river or getting ground down by a glacier. When plants arrived, they brought with them a whole new suite of ways to create mud. Their roots exuded sugars and acids that crumbled bedrock into silts and clays, through an erosive process called chemical weathering. White webs of fungal mycorrhizae that grew in association with roots released organic compounds of their own that penetrated stone and opened up wounds in rocks that led to yet more chemical weathering.

As rocks chemically weather, they react with carbon dioxide dissolved in rainwater and pull this greenhouse gas out of the atmosphere. Over time, this process can have a cooling effect on global climate. If the exponential rise of mud represented a surge in plant-induced chemical weathering, then this would reveal that plants had a stronger sway on this part of the carbon cycle than previously thought. If the rise of mud, instead, represented more of a retention of existing mud rather than the creation of new mud through chemical weathering, this would reveal that plants played a slightly

different role in the carbon cycle. As climate modelers work to predict future conditions, they need to understand these fine details
of the carbon cycle and—importantly—how they play out over
geologic timescales.

To find answers to their questions, Neil and Will are collecting
mudstone samples from around the world to bring back to the lab
to tease out their origins. Beneath strong enough microscopes, they
can distinguish chemically weathered silts and clays from physically
weathered ones. Through these analyses, they hope to triangulate
how the muckier way of things came to be and how, exactly, it
changed life on Earth then and thereafter.

Nine

IRELAND'S DINGLE PENINSULA DANGLES LIKE A BIG toe off County Kerry's southwestern coast, in a tangle of gorse, holly, willow, and ferns. Leaves unfurl from gutters, trunks sprout from broken windows, flowers rise from old stone bridges, and dense hedgerows knit together miles of pasture that carpet the ground in green.

The peninsula's only town, called Dingle Town, sits at the edge of a turquoise sea. Gem-toned stucco buildings line narrow streets that lead up into rocky hills that overlook a quiet harbor. Even with such striking scenery, it's the vegetation that steals the show here, glowing almost neon green when the sun peeks out from behind clouds, winking at the nickname Emerald Isle.

"It's what we're known for," a woman told me on my bus ride from Dublin.

It seems fitting that a region so defined by its lushness should contain some of the world's oldest traces of plants that grew on land.

I went there not so much for those early plants, but for the mud laid down as a result of them. I arranged to drive around the peninsula with Neil Davies to look at rocks from the Silurian (443.8 and 419.2 million years ago) and the Devonian (419.2 and 358.9 million years ago), which were the first two geologic periods when land plants grew widespread across the continents.

I arrived on the peninsula a day early to get settled and orient myself. The symbol of a castle on my tourist map sent

MUD

145

me to a rocky beach a few miles outside of town where I found the sixteenth-century ruins of Minard Castle. The structure did not have much in the way of turrets or other flair that the icon on my map had led me to imagine. It stood tall and square, flaunting a handful of narrow windows that looked out over a moody ocean. A fence prevented visitors from approaching the ruins close enough to touch, so a group of us stood at a distance, some lifting binoculars to their eyes, others complaining about not being able to get nearer.

I was a little disappointed myself. I had wanted to step inside and look up through a stone window that someone else had built hundreds of years earlier. I had wanted to imagine their hands and their clothes and their thoughts as they looked out across the same ocean that I looked out at on that Wednesday in the twenty-first century. I had wanted, in certain ways, the same thing that I want from strata, just on a human scale.

Though we couldn't touch the castle, we could touch the rocks that sat beyond it on the far side of the beach. I strapped on my backpack and boulder-hopped toward those cliffs instead. On my way, I passed an informational sign that, to my surprise, wasn't about the castle at all, but about the geology of those rocks I was heading toward. The sign makers seemed to have shared my train of thought: *The castle is cool, but look all around you.*

The captions on the sign explained that the rocks in the cliffs and in the castle had formed more than 300 million years earlier, when Ireland sat south of the equator and plants and mud were just beginning to cloak continents. This particular spot had been covered in desert dunes at the time but, elsewhere nearby, land was beginning to green and muck up.

I approached the cliff of red and ochre strata and found the same types of swooping herring bone layers that Ryan and I had seen in the copper mine in South Australia. I got nose close to the rock and found rhythmic pinstripe patterns of coarse sands and fine silts, followed by more layers of sands and silts. Each finger-thick couplet represented gusts that drove the migration of this dune steadily in the direction of the wind.

Encountering such familiar couplets in this place I had never been felt like overhearing a familiar language spoken in a far-flung part of the world, or finding a familiar constellation in an unfamiliar sky. I dragged my thumb up and down the layers and felt grounded by each pulse, each quiet avalanche that had pushed the dune forward so long ago. They reminded me how this cyclicity, this giving into gravity to make way for something new, has always existed at the center of the Earth system, at the center of each life.

Small pink flowers growing out of a crack in the rock released a sweet, buttery aroma that relaxed my lungs. Their rustling stems whistled in my ear. The arrival of plants hadn't just changed landscapes, but smellscapes and colorscapes and soundscapes too.

I sat on the cliff for a while, watching the castle made of dunes made of sand slowly crumble, grain by grain, into the ocean. The tide sloshed backward and forward, reshuffling bits of the story and shaping it into something new.

◆ ◆ ◆ ◆

NEIL DAVIES SPEAKS IN a low voice and answers questions graciously but without flourish. As a kid growing up in Staffordshire, England, he spent his days hiking, caving, climbing, and strolling the limestones and sandstones of the United Kingdom. It was this preference for spending time outside that eventually led him to geology and, in turn, to mud.

Though he can be tight-lipped at times, he does offer up some opinions when prodded. He dislikes fashion, for example, and also when scientists use words like "groundbreaking" or "exciting" to describe their research.

"I prefer to choose my own adjectives," he tells me. It's not that he's a grinch so much that he worries that such superlatives cloud the actual merit and rigor of scientific findings. Or, worse yet, pull

M U D 147

attention away from less glitzy but equally important findings in the rock record. As someone who studies grains of silt and clay for a living, he would know.

"You can't appreciate what's special," he says, "without appreciating what's boring."

It's a tenet of his research that he'd repeat throughout our time in Dingle, a mindset that he considers integral to stratigraphy—a mantra that could just as easily describe the path to Buddhist enlightenment. Only by emptying our minds and appreciating the seemingly mundane can we rise out of our own egos and see and feel what really matters in this world.

Though we generally use the word *mundane* to describe things we may find boring, the word has roots in *mundus*, the Latin term for "world," as in, of this world rather than of the heavens. This duality of meaning—of this world and boring—perfectly embodies the fundamental nature of stratigraphy. It requires a commitment to paying attention to the events that have unfolded across this planet millions of times over. A commitment to patience, to breathing in and out and finding meaning in the ordinary.

Without an appreciation for the mundane, someone like James Hutton—the "Father of Modern Geology"—may never have heeded his gut feelings of there being "no vestige of a beginning,—no prospect of an end." Westerners might still be wandering around thinking that all rocks visible on Earth were laid down in a relatively recent cataclysmic flood, rather than through unimaginably long expanses of not much happening at all.

When Neil and two graduate students found what turned out to be the world's largest millipede fossil on the coast of England in 2019—a behemoth that measured the length of a VW Bug—he left a piece of the specimen sitting on a swivel chair in his office for three years before doing anything with it.

"It had been in a cliff for 326 million years," he reasoned. "What's three years in an office chair to it?"

The find, while certainly fun and exciting (my words, not his), added relatively little to the understanding of how these animals lived. Fossils of the same species had cropped up before, just none quite as big as this one. Still, the discovery eventually made headlines when he finally published a study on it in 2021. "Suddenly there is an article about a giant millipede that is not about war or plague or politics and people kind of like it," he says. On social media, someone mentioned overhearing two lovebirds talking about the millipede while waiting for a bus in Jakarta. "It makes people happy," Neil concedes. "And that's good."

"And people could be more happy," he goes on, "if they spent more time looking at rocks." You won't always find giant millipedes, but you might find evidence of the first microbes that puffed oxygen in the atmosphere, or an ice age that covered the entire planet. You'll find infinite sources of awe and humility. "You'll get to appreciate that there's big stuff back there in deep time."

More often than not, though, you'll find the mundane, daily occurrences that have been sloshing around this planet since the beginning. The remnants of gusts of wind, periodic floods, tides washing over beaches.

To train his eye to differentiate these mundane moments from more anomalous events, Neil spends as much time in the field as possible, looking at rocks laid down across many different times in Earth history. When we met, he was forty-two years old and estimated he had spent a cumulative three years of his life out in the field—more than most geologists do in their entire careers.

"He's like a dog who needs walking," says Will McMahon, who moved on to work alongside him as a postdoctoral researcher after completing his PhD project on the rise of mud.

Neil hasn't spent quite as many months in the field as someone like Paul Hoffman but, at half Hoffman's age, he has plenty of time to catch up. Some field geologists might let such experience go to their heads, but Neil's time with the strata seems to have had the opposite effect.

MUD

149

"Field geology should dampen any concept of ego," he tells me. "All that stuff, all that deep Earth history, and then you're around for what, seventy years? And you think you're special? You're not fucking special."

Nobody is, he says. At least not in the scope of geologic time.

✦ ✦ ✦ ✦

WHEN I MET NEIL in downtown Dingle, he towered above my short frame as he introduced me to his new graduate student, Yorick Veenma, who would be joining us on our road trip around the peninsula. The two of them ducked into a nearby supermarket to grab sandwiches, donuts, and a box of ziplock bags for sample collection, and then led the way to their rental car to begin our journey through time. They had arrived earlier in the week to meet up with a group of Swedish colleagues and, based on the pungent pile of daypacks in the back seat of the car, I understood that it had been an active few days.

Once we escaped the hubbub of downtown, I asked what seemed like a basic question, but one that got at the heart of Neil's research: If plants had never risen up on land, how would this place look different?

"Nothing here would be the same," Neil said, glancing his dark blue eyes into the rearview mirror and shifting gears from the driver's seat.

Fair enough. The Emerald Isle would no longer be emerald. In my mind's eye, I removed the cows munching on the pastures we sped past and then removed the pasture itself. I removed the mud underneath the pasture and stripped the land all the way down to bare rock, narrating my thought experiment aloud as I went.

But even that, Neil pointed out, would be different. The rock was made up of sands and muds deposited in a world manipulated by vegetation. If that flora had never existed, those rocks wouldn't have formed the way that they had. "It's not just these living plants, but the immediately dead ancestors that make up the soil, and the deep time ancestors that are making up the rocks."

Plants didn't only influence the contents of the bedrock, but the shape of it as well. The glaciers that retreated at the end of the last ice age some 10,000 years ago may have carved through the land differently or failed to take shape at all if plants and mud hadn't been around to add complexity to the climate system.

Before plants and mud arrived, the carbon cycle was far simpler. Carbon dioxide arose in the atmosphere mostly by sizzling out of the mantle through volcanic activity, and slinked out of the atmosphere primarily by dissolving in rainwater and getting trapped in minerals on the seafloor. Other carbon sources and sinks existed, but the non-living components of the cycle held more sway.

The arrival of plants and mud threw new loops into this cycle that drives global climate.

Plants also introduced subterranean piping that brought water seeping underground in volumes that would have been previously impossible. Today, an acre of prairie may contain as much as 24,000 pounds of water-pumping roots below ground. That water flows down through roots and evaporates back out through the tips of leaves, blurring the boundary between ground and air.

"It's not really clear what is earth and what is sky," Neil said.

Yorick sat mostly listening as we spoke. His studies were focused more on the distribution of mud in different environments across the planet *before* plants arrived on land, to provide some context for the conditions plants evolved into.

When I asked what drew him to this particular facet of geology, he said it was his fascination with the idea that life can behave as a geological force—not just plants but also animals and fungi and bacteria. All living beings. "Life evolves and the environment evolves with it," he said.

Since the very first cells coalesced, life has been tweaking the distribution of elements and minerals and gases that in turn shape the

MUD

physicality of the planet. Living things have never been passive actors on a stagnant stage, but rather active sculptors of the very material that they spring from. This continues to be true to this day, both in obvious ways, like beavers building dams, and in far less obvious ways. Even something as seemingly irrelevant as the placement of Pacific salmon eggs in streambeds has been shown to have the power to dictate the shape of surrounding mountain belts, according to a 2017 study titled "Sex That Moves Mountains."

We sat for a moment in silence, looking out the car windows at cows lazily chewing on pasture, standing against a backdrop of hills carved by glaciations orchestrated by plants and mud.

Ten

IN THE ROUGHLY 300,000 YEARS THAT *HOMO SAPIENS* have roamed Earth, Charles Darwin could not have been the first person to consider how life shapes and reshapes landscapes over time. But he was the first to write a bestselling book on this notion, a treatise on earthworms that he published in 1881, the year before he passed away.

Through a series of observations across Great Britain, Darwin calculated that earthworm populations consumed, loosened, and mixed as much as ten tons of soil in a given acre of land in just one year. That's enough dirt to outweigh an elephant.

"The result for a country the size of Great Britain," Darwin wrote, "within a period not very long in a geological sense, such as a million years, cannot be insignificant."

Within that geologic timeframe, he estimated that earthworms would pass as much as 320 trillion tons of earth through their bodies—more than one million times the combined weight of all the elephants alive on the planet today.

Since Darwin published those estimations, the field of biogeomorphology—the study of how living things morph the landscapes they inhabit—has revealed the countless other ways that life shapes the surface of the planet, from the tiniest of living things to the largest.

In the case of mud and plants, the biogeomorphology plot twist of note is the change in river shape from multichannel braided streams—those that flowed loose and chaotic like rivulets at the edge of a beach—to single-channel meandering

MUD

153

ones like parts of the Amazon and Mississippi. The term *meandering* here refers to the sinuous, snakelike form that single-threaded rivers assume as they migrate across a landscape. Because braided rivers lack cohesion, they can't meander or snake but, instead, cut straight across the land they flow through.

While braided rivers became less common as plants and mud added cohesion to the landscape, they still flowed in the newly vegetating world, and they continue to flow on the modern planet as well. The classification of rivers is not so much a binary system as a spectrum of shapes and habits. Parts of a single river can braid and then meander and then braid again at the mouth. The slope of the land, the amount of water and sediment suspended within it, and the extent of vegetation growing along banks all contribute to a river's classification. Today, braided rivers tend to form at the base of mountains and glaciers—locations with high volumes of both water and sediment, where channels continually spill over their banks and rewrite their routes again and again to the point that the banks become barely definable at all.

As the amount of sediment and volume of water wanes downriver or over time, that braided channel will naturally evolve into a sinuous meandering one. It will narrow as it drains, like water in a bathtub, and vegetation will opportunistically take hold and further reinforce the cohesion that allowed it to take a sinuous shape in the first place. That vegetation will add drag and reduce the speed of the flow, giving fine silts and clays the chance to settle out and further narrow the channel as these particles accumulate as sticky mud.

Pinpointing how and why the shapes of rivers changed hundreds of millions of years ago may feel far removed from modern-day concerns. But studies of ancient river deposits from before and after the greening and muddying of the continents have helped clarify our understanding of modern river systems and how we might better manage those systems to support the humans and wildlife that live among them.

"If you don't understand what's driving the river into one state or another, it's hard to do that well," says Chris Paola, a sedimentologist at the University of Minnesota.

Take the Platte River, as an example. This naturally braided tributary of the Missouri River runs across the entirety of Nebraska and offers roosting and feeding grounds for a number of migratory birds, including endangered whooping cranes. With wingspans that stretch up to eight feet wide, and standing at heights that tower above most flamingos, these elegant white birds are North America's tallest. They once flew in great numbers across the continent during their annual journeys from the Gulf Coast to Canada, but have suffered from habitat loss over the past century. Historically, they would sleep atop sandbars within the Platte's shallow channels and feed within the surrounding prairie by day during their several-week stopover. But as damming and irrigation have reduced the frequency and magnitude of floods downstream, more vegetation has taken hold along its banks and has forced the river to behave more as a meandering single-threaded river than a braided one. Its channel has responded by narrowing, deepening, and flooding the sandbars that the cranes had once slept on, leaving them nowhere to safely rest out of reach of predators.

"From a habitat perspective, this was a disaster," says Michal Tal, a geomorphologist who consults for the Platte River Recovery Implementation Program, which is working to restore critical habitat by encouraging the river to morph back into its naturally braided state.

In order to solve such human-made problems, river managers must have some concept of the forces at play in the evolution of a river's shape. But because rivers shift course and evolve new shapes over the span of decades, rather than the few years typically allotted to management plans, it can be difficult for managers to develop effective long-term plans. Many overlapping variables often impact a river at the same time, making it difficult to isolate the individual effects of different variables—another reason experiments in controlled settings prove useful.

Michal was aware of this conundrum when she enrolled as an

Earth science graduate student with Chris Paola at the University of Minnesota in the early 2000s, and committed her PhD research to trying to develop tools to help resolve it. If, in a laboratory setting, she could speed up and replicate the evolution of a braided stream in response to vegetation growing along its banks, she reasoned that she could pinpoint some of the forces and factors at play in the real world.

Michal knew that other research groups had attempted to reproduce meandering channels in the lab, and that they all had largely failed to create the type of self-maintaining channel that exists in nature. Those other experimental streams all devolved into disorganized systems with multiple channels, unable to maintain their cohesive sinuous shapes over the course of multiple flood cycles. All the failed attempts had one thing in common: they lacked vegetation. But she and her research group were well aware of the stratigraphic record and what it indicated about the relationship between plants and meandering streams during the Silurian and Devonian— how, as plants rose up and multiplied across the continents, so too did the abundance of sinuous meandering channels. "That was the missing ingredient," Michal told me.

Inspired by the insights from the rock record, she got to work designing her experiment, which followed on the heels of the work of a previous student. In the windowless hum of the St. Anthony Falls Laboratory in Minneapolis, she filled a sixteen-meter-long steel flume with uniform white sand marketed for swimming pool filters. She tilted the trough at a slight angle, carved a shallow depression in the sand down the length of the flume, and cranked on a flow of water. Within a few hours, the water had organized itself into a series of channels that drifted around numerous sandbars. She had created a classic braided stream.

From there, Michal increased the water discharge and let loose a "flood," watching as miniature sandbars eroded away and formed anew elsewhere downstream. After an hour, she reduced the flow to a trickle to end the flood. She threw handfuls of alfalfa seeds across the sand and left the system running on this low-flow trickle for several days, emulating patterns that might take place over the course of a year within a natural temperate river system.

She repeated this "annual cycle" of low flow and floods every five days or so for about five months, generating "decades" of river evolution. Over that time, the flume channel took on a life of its own. It migrated, eroded its banks, developed new channels, and abandoned old ones as floods came and went and as vegetation began to crowd in. "It was just like watching a real river," Michal told me. "It was totally self-evolved." What began as a wide, shallow, sandbar-laden braided channel steadily morphed into a narrow, deep, sandbar-less single-threaded channel. The alfalfa had taken root and stabilized the sediments.

She repeated the experiment with slight variations three more times and recorded the results with time-lapse videos that she condensed down to less than a minute of footage.

"When we showed those videos to river managers," Michal told me, "their mouths dropped." Just as she had hoped, her experiment proved valuable for those working to design and justify management plans in modern braided rivers. The Platte River Recovery Implementation Program had already begun buying up miles of land along the river to bulldoze away vegetation to help maintain its braided nature, and had been doing timed releases of floodwater to prevent the establishment of plants in the first place. Now they had experimental proof that these actions should keep the river braided in the long run, even if there were periodic anomalies in the system from one year to the next. This proof helped them garner support for restoration practices that might have otherwise faced more pushback.

On the flip side, the videos also proved helpful for Earth historians such as Neil Davies, validating their observations in the rock record and providing a mechanism to explain them. The present is the key to the past, as Charles Lyell's maxim goes in *Principles of Geology*, the book that helped Darwin come to grips with geologic time.

MUD

But the past is also key to the present, and we need researchers working in both realms to bridge the divide between them.

✦ ✦ ✦ ✦

AFTER DRIVING WITH NEIL and Yorick down a narrow coastal highway for a few dozen miles, we pulled over at Slea Head, a scenic promontory on the southwestern tip of the peninsula. Famous for its bleached white crucifix that solemnly faces out across the North Atlantic, this bend in the byway swarmed with American tourists collecting photos on their phones. Neil executed a series of tight maneuvers around them in what was left of the narrow pullout, and we stepped out into a torrent of wind. Yorick stuffed ziplock sample bags into a large fanny pack, Neil piled his gear into a khaki satchel strung across his chest, and we all crossed the road together and headed not toward Jesus but to the rock just beyond him.

Neil and Yorick would be taking advantage of this stop to collect samples to bring back to Will, who was examining the microscopic structures of rocks like these to determine what proportion of mud had been generated by plants through chemical weathering at this time. The stop also offered a bit of geologic sightseeing on my behalf to help illustrate evidence of biogeomorphology in the rock record.

Neil had chosen this particular location as our first stop based on a geologic field guide that identified the rocks as remnants of a braided river that flowed early on in the Devonian, when plants and mud were still new to continents. The guidebook also mentioned the possibility of trace fossils—burrows left behind by something like a millipede—so we'd try to find evidence of these early land dwellers too.

As the other tourists chattered around us and snapped pictures on their phones, we laid our hands on the grayish-green and maroon-tinted strata, quieted our minds, and turned inward.

If you follow other types of scientists into the field, you'll find it's often go, go, go. It's searching for that one bird or chasing that one eruption or motoring the boat upriver to collect that one water sample. In stratigraphy, the pace reverses. It's stop, stop, stop. Look at what's in front of you. Let the rocks wash over you. Let the stories speak for themselves and take form in your mind's eye. Let them inhabit you until you're not on a roadside in Ireland but on an arid continent beneath the equator, a river flooding before you, and nothing much larger than your face growing on land.

To an onlooker, this might not look like much at all. Pencils behind ears, hands on hips, eyes on rock. Steady breathing in and out. But telescope into the mind of the stratigrapher and that onlooker would find a river unspooling and surging over cobbles and pebbles, rearranging itself through mazes of sandbars.

A car whizzed by, jarring me back to the present. I hugged the shoulder of the road and asked Neil to narrate what he saw in the mess of sediments in front of us in the rock. The grains ranged in size from grapes to grapefruits and rested within a matrix of green-gray sand. Based on a few quick observations, he recognized this as a river deposit and not a snapshot of the deep sea or a sand dune or a mudslide. For one, the fruit-sized pebbles and cobbles were too large and heavy to have tumbled all the way down to the deep sea. They were also too large to have been transported by wind, which generally only moves poppyseed-sized grains like the ones visible in the dune deposits alongside Minard Castle.

A mudslide would have been strong enough to move the pebbles and cobbles, but would not have left them as neatly organized as they appeared in this outcrop.

To illustrate this last point, Neil grazed his index finger along several slender oval pebbles and pointed out how they all aligned vertically and tilted in the same direction, like books collapsing on a bookshelf. "Once things roll around inside the system," he told me, "they are going to self-organize in a way in which it's easier for water to carry on flowing around them."

I imagined a strong current roaring over those stones, arranging

MUD

them in the midst of a flood. Unlike the randomly plopped dropstones in the glacial marine sediments of Svalbard, these had a uniform direction to them.

I squinted my eyes and began to notice more organized bands of cobbles and pebbles, separated by bands of finer-grained sediments. Generally speaking, the larger the grain in a riverbed, the stronger the current that deposited it.

I approached one lighter-colored band and found that the grains were smaller than sand. They were a cohesive collection of silts and clays. Mud.

I dragged my finger across the mud and was surprised by the dusty residue it left on my skin. Even after all of those hundreds of millions of years, it hadn't fully solidified. I held my finger beneath my nose, breathed in, and arced back through time by way of the faint dank smell of a riverbed.

I pointed to a band of pebbles and cobbles that measured the length of my body and asked Neil how long it might have taken to accumulate. He scanned it for a moment and then spat out an estimation. "Literally seconds," he said. At least for the few pebbles right in front of us. The entire band may have taken minutes or hours.

Hours of chaos followed by who knows how long of not much happening at all. This is what makes the geologic record and geologic time so challenging to comprehend when studying it in outcrops like this. You have to be just as cognizant of what *wasn't* happening as what was, and remember that what was and wasn't happening could have been taking place for anywhere between seconds to hundreds of millions of years.

The rock record is tricky like that, Neil told me. "It shouldn't be surprising that large amounts of it are understandable on the human timescale, but also that it actually encompasses a ridiculously long period of time as well," he said. "You have to find ways to understand it, speak its own language."

He wasn't convinced we'd find the trace fossils mentioned in the book, and muttered something about the authors of the field guide being metamorphic rather than sedimentary geologists—a classic jab sedimentary geologists make at subdisciplines that study layers that form deeper below Earth's stratified crust. The absence of fossils did not deter Yorick from knocking off a handful of samples to bring home with him, as the mud was more relevant to their research than the fossils.

We returned to Jesus and made our way down the road to travel further back in time.

+ + + +

WHEN NEIL DAVIES AND Will McMahon published their findings on the rise of mud in 2018, many of their colleagues were struck by the findings.

They weren't particularly surprised that mud's accumulation coincided with the rise of land plants, since that had long been assumed from the rock record. But they *were* surprised that the rise of mud came *before* the rise of roots and stems—materials long thought to be responsible for trapping silts and clays. According to their findings, even these earliest rootless growths (including one of the planet's earliest mosses, *Dollyphyton boucotii*, named after Dolly Parton) helped trap some silts and clays, and this effect only intensified as plants grew larger and leafier and more capable of gathering debris. By the time plants evolved to be tree-sized, they developed the capacity to slow wind and force silts to drop to the ground when gusts died out in branches. Tree canopies and leaf litter reduced the impact of falling raindrops, further easing erosion and helping banks gather more muck.

But could the earliest of land plants without roots and stems— those diminutive ancestors of mosses and liverworts—really have added enough roughness to the landscape to trap orders of magnitude more mud on land?

Like Neil, Woody Fischer jumps around between different epochs in Earth's past. Though his research focuses largely on

MUD 161

the Great Oxygenation Event, he dabbles in other events as well, including the rise of mud on land. When he and his colleagues at Caltech came across Neil and Will's findings, they began thinking through possible explanations for how the earliest plants may have instigated mud's terrestrial ascent. They understood that mud's stickiness isn't just a physical quality, but a chemical one with traces all the way down to the molecular level. This chemical quality, they reasoned, may also have had something to do with clay and mud's accumulation.

If you were to look at a single particle of clay beneath a strong enough microscope, you'd find a tiny stack of flat mineral sheets all piled on top of one another, like the flaky phyllo dough layers of spanakopita. Clay minerals are, in fact, part of a larger mineral group that shares the same etymological root as this Greek pastry— the phyllosilicates. Their mineral layers cling together not with strokes of melted butter but with electrically charged particles. It's this built-in charge that makes clay minerals so sticky, adhesive, and good at trapping carbon underground.

Sarah Zeichner, a former graduate student of Woody's, has spent a lot of time thinking about phyllosilicates and how they behave in the environment. She takes the food metaphor one step beyond phyllo dough and likens clay particles to sandwiches lathered with sticky charged "condiments" that hover within the empty space between the silicate slices of "bread."

"You have a place to start trapping your cheese or your sprouts, or whatever you want to put on your sandwich," Sarah says. These ingredients can include other silts and clays, or organic detritus like the bits of dead plants or microbes or animals, which the clay particles keep trapped away from decomposers.

Geoscientists have long debated the role clay plays in the carbon cycle by trapping organic material and storing it underground. But the flip side of this relationship—the capacity of organic material to help trap clay on land—hasn't received as much attention.

When Sarah, Woody, and others in their research group began

mulling over Neil and Will's paper, they wondered if this other side of the equation might have played a role in the unexpected timing of mud's rise. Because they are so lightweight, clay particles on their own struggle to settle in moving freshwater. The slightest river current will churn them up and pull them all the way out to sea. But if circumstances allow them to gather together into larger globs, called flocs, they become heavy enough to sink.

Much in the way that specks of dust and pollen give snowflakes something to grow around, specks of organic debris in water can help clay gather together into sinkable flocs, through a process called flocculation. Prior to the rise of plants on land, such organic debris would have been largely absent from bodies of freshwater. Dead fungi, bacteria, and perhaps some algae would have floated around to a certain extent, but likely in underwhelming quantities. As the continents greened, however, the amount of material available to spur flocculation would have exploded. Maybe, the Caltech researchers thought, this was how mosses and liverworts contributed to the earliest stage of mud's accumulation on land, before the stabilizing power of stems and roots came into the picture.

To test this idea, one researcher in the lab threw together a bunch of beakers with clay and began conducting some loose experiments. But before long, he took a job elsewhere and left the jars to gather dust on the lab bench.

Sarah, as the curious and enterprising first-year graduate student that she was at the time, jumped in and began designing a set of experiments of her own. She ordered bags of pure mineral kaolinite clay from her lab-supply store and filled the beakers with purified water. She then placed the beakers full of clay and water on magnetic plates, plopped a small magnetic bar in the bottom of each, and flipped a switch to make the bar spin and swirl the water, creating something like a river eddy.

The final thing she needed for the experiment was a material to emulate plant detritus. She couldn't simply grab a pile of dead jacaranda leaves from the campus courtyard and throw them in the beakers because that would introduce too many variables and unknowns. She needed some sort of pure and uniform source of organic debris to match her pure and uniform clay and water.

In brainstorming where she might find such a thing, her mind went, again, to food. She pulled up a handful of YouTube food science videos, watching studiously as chefs whipped liquid fruit juice into solid vegan caviars and gelatins, with the help of simple organic binders. She hurried to her local organic food store and found some of those same ingredients in the baking aisle: bags of Bob's Red Mill xanthan gum and guar gum. These ingredients may sound like synthetic concoctions, but they actually come directly from the environment. Xanthan gum comes from a strain of bacterium and guar gum from the seeds of a bean native to India.

Sarah brought her groceries back to her lab bench and added pinches of them to her beakers. Wanting to prove herself to be a serious scientist, she also got to work pulling together a computer model to help quantify her observations. Over the next week, she typed away at her model and clinked around the beakers. But, in the end, the hustle and the computer model weren't even necessary to prove her point; the results were clear to the naked eye. Without the added ingredients, the clay remained in suspension for days at a time. But with a small pinch of "detritus"—just 0.1 percent of the weight of the clay—the particles glommed together as flocs and sank within a matter of seconds. "That was shocking," she told me.

The organic debris had, indeed, spurred flocculation, and had done so with far more vigor than she and her lab mates had expected. With the help of a high school student she was mentoring at the time, she found that the guar gum—the plant-based ingredient—pulled the clay grains together far faster than the bacteria-based ingredient. This suggested that the arrival of plants on land could have helped clays flocculate more readily than they would have just in the presence of microbes. This could at least partially explain the timing of the rise of mud before the rise of roots and stems, and she reported as much in a paper that she copublished in *Science*.

164   STRATA

While a beaker full of clay and guar gum and a stir bar in a lab isn't a river, it offered a reasonable way to tackle some of mud's unknowns. The strata in Dingle would help fill in more pieces of the story.

✦ ✦ ✦ ✦

AFTER A BRIEF PIT STOP at Europe's westernmost public bathroom, we hugged the coastline north to a sandy beach surrounded by neatly cropped sheep pasture. Neil estimated that the rocks there formed some 20 million years earlier than the ones at our previous stop, placing them squarely within the Silurian when vegetated floodplains were just beginning to take form alongside rivers.

We parked in a small lot facing out onto turquoise waves that crashed angrily ashore. The power of the swell felt discordant with the soft breeze ruffling our hair. "They have come from North America somewhere and are just taking it out on this beachfront car park," Neil said as I stared perplexed at the oversized waves. My mind's eye zoomed out and saw how the swells had gathered force not from local weather patterns but from momentum gathered across thousands of uninterrupted miles. They connected us back to my home on the other side of the Atlantic, where the winds that shaped them first whipped offshore.

With our bags packed up again, we trekked across the beach to the rocks just beyond it. Yorick ducked away with hammer in hand in search of mudstone while I stayed behind with Neil to look for evidence of the ancient vegetated floodplain that he believed we were standing on.

More often than not, it's the chemical ghost of an ancient root rather than the plant itself that gets left behind in the rock record. Plants, after all, tend to decompose. Conditions have to line up just right for fully intact vegetation to become buried in ways that preserve their structure, and still further conditions have to line up just right for those deposits to turn into sedimentary rock and then rise back to the planet's surface somewhere accessible and visible to humans. It's not impossible for everything to line up perfectly, but it's not particularly common either.

The chemical traces of roots, however, are far more plentiful. Whereas an inert object like a shell or a bone will generally only take up physical space or leave behind a physical imprint in the rock record, roots have the power to chemically manipulate the ground they grow within. In so doing, they leave behind a more complex geologic legacy. As they go about their daily chores of feeding and hydrating stems and leaves, roots exude cocktails of sugars and acids from their tips that change the makeup of the sediments and the underlying bedrock. Roots also support microbial communities that leave behind visible chemical traces.

Within a few moments of poking around some seaweed, Neil called me over to where he stood and pointed out one such sign of ancient vegetation in a piece of reddish-gray stone that had chipped off the bedrock beneath our feet.

"Calcrete," he said, pointing to a mangled band of white chalky material stuck within the stone. This white mineral growth, he explained, forms within soils that cycle through periods of wet and dry over long stretches of time. Water seeps in and leaves behind the white residue as the soils dry, like lime at the bottom of a tea kettle. The finger-thick chunk of calcrete that Neil held in his hand must have formed within ground that cycled through many periods of wet and dry. Like, say, a vegetated floodplain, with its periodic bursts of water and its stabilizing muds and roots. These are the types of clues he looks for to reconstruct ancient vegetated environments.

We continued on with our search and, before long, found a piece of bedrock with a striking leopard-print pattern of red and pale gray spots. Ferricrete. Like calcretes, ferricretes develop in soils that cycle through periods of wet and dry. The dry periods rust iron within the rock and turn it red (ferri-, iron), while wet periods leave the iron a pale gray color. As with calcretes (calc-, calcium), the waterlogged conditions of floodplains help form ferricretes, and plants help form those waterlogged conditions. These were still not definitive signs of a vegetated floodplain, but it felt like we were getting warmer.

Finally, Neil picked up a loose piece of ferricrete with a pale upside-down Y-shaped blotch that appeared tattooed straight through it. Bingo. This, Neil told me, was the chemical halo that would have formed around a root as it pumped out its various organic chemicals. I placed my finger on the edge of the Y and imagined the lacy tendrils that would have reached down and out into the newly lush world. The notion of trees and flowers and fruits was still millions of years in the future; for the time being, these roots were the most sophisticated invention of the vegetable world.

When Yorick returned, he hauled with him a generous collection of mudrock, further confirming our hunch that we were standing on an ancient muddy floodplain.

As we made our way back to the car, Yorick wandered off again and came back beaming with what looked like a hand-sized chunk of fossilized corals. Neil's face lit up as he examined the sample. "You can't see inside a volcano or inside the middle of the earth, but you can see a coral there, and you know what a coral looks like," he said, handing it to me. Such recognizable forms provide a tangible connection with Earth's past. "It's like a little passport to deep time," Neil said.

Eleven

THE WESTERN WORLD'S UNDERSTANDING OF DEEP time has roots not far from Dingle, along the eastern shores of Scotland. It was there that James Hutton made the observations that would seal his fate as the Father of Modern Geology and forever change the way that Westerners understood the depth of the geologic record.

Hutton spent the years following his famous 1785 lecture to the Royal Society of Edinburgh searching for the perfect outcrop to help drive his points home. Until then, his theories had come mostly from his mind and from observations of modern geologic processes taking place on his farm. He needed more concrete evidence in rocks to convince naysayers of what he knew in his gut to be true—that landmasses have built up and broken down and built up again, over and over, on and on, not since forever but since long before the 6,000-some-odd years allotted to Earth within the Bible.

Hutton continued searching for that perfect outcrop for several orbits around the sun, looking for a single location that illustrated incremental and cyclical changes in a landscape through time. One with layers that formed in demonstrably different environments, stacked on top of one another like pages in a flip-book.

As the years spun on, he continued to come up short of what he was looking for. By that time he was in his early sixties and could no longer easily explore the rugged coastline of Scotland by foot. Instead, he accepted the help of a younger friend, chemist James Hall, who offered up a boat and a waterfront

168   STRATA

property from which to launch it. In June 1788, they set sail together with their mathematician friend John Playfair and a few farmhands in search of the perfect outcrop by way of the sea.

Carried by the currents of the North Atlantic, they approached a number of rocky points, each time buoyed by optimism that sank to disappointment as steadily as the waves rose and fell around them. Finally, they rounded a bend and came across a jut of land called Siccar Point. The rocks there sent Hutton into a tizzy. His crew rushed the boat ashore and scrambled over a rocky beach to get a closer look.

It would take Hutton's narration to bring the seemingly mundane rocks to life. He recognized the base of the outcrop as a type of flaky, coarse-grained metamorphosed stone composed of seafloor sediments that he called schistus (but that is now known to be greywacke). He knew that those layers must have originally sat flat, since that's the only way that sediments can accumulate on the seafloor. But there at Siccar Point, the layers sat vertically, like books upright on a bookshelf. Some geologic force must have pushed them on their side to lay them that way—and that must have happened only once the layers had solidified into stone, a process that itself must take longer than the human mind can comprehend.

On top of the gray vertical stacks of rock sat a thinner deposit of coarse sands and pebbles that were composed of the same material as the gray rock below. They appeared to have broken off that ancient seafloor once it had turned to stone. This, Hutton recognized, was evidence of the erosive forces of wind, rain, ice, or snow—evidence that what once had been a seafloor had risen up above sea level and sat out in the open air, perhaps along a mountainside.

This may sound like a lot of information to pull from a few layered rocks, but the beauty of stratigraphy is the ease with which you can make such interpretations once you learn the basic alphabet of the sediments. Layers can only accumulate horizontally in bodies of water. Rocks break into smaller bits when exposed to open air. Hutton understood these mundane concepts simply by observing them playing out in real time on his farm.

M U D   169

From his years of observation, Hutton understood that the time it would take to turn a seafloor into a mountaintop that later eroded down into smaller fragments would transcend the time allotted within the Bible.

But the signs of Earth's antiquity did not stop there.

Atop that deposit of pebbles sat a thick package of horizontally layered red sandstone. The seafloor that had turned into a mountaintop must have turned back into a seafloor where these horizontal stacks of strata accumulated. The yo-yoing of geologic processes must have then sprung the package of stone back up above sea level where those men stood that day.

There was nothing superficially breathtaking about the rocks that the North Atlantic splashed against on Siccar Point. The deposits looked similar, in many ways, to the rest of the coast of the British Isles. By certain standards, they might be considered boring. But Hutton, like Neil Davies, knew to look beyond the surface of the mundane. Whereas those before him called upon a dramatic, cataclysmic flood to explain the rocks we live among, he called upon the imperceptibly slow and steady movements that continue to play out across the earth today.

"For Hutton," writes biographer Jack Repcheck, "there was no need to call upon unseen and unknowable catastrophes from the past. . . . The inexorable forces of wind and rain, tide and waves, volcanoes and earthquakes, which the earth still experiences every day, formed the world we inhabit. All that was required, as Hutton stated, was 'immense time.'"

As Hutton narrated the scene to his compatriots at Siccar Point, something finally shifted. Both Playfair and Hall had grown up as religious men and had been hesitant to subscribe to Hutton's ideas early on in their friendship. But over years of conversation and consideration, Repcheck writes, they began to warm up to the concept of deep time. Siccar Point sealed the deal.

"On us who saw these phenomena for the first time, the impression made will not easily be forgotten . . . ," wrote Playfair in an

1805 account of the discovery. "We felt ourselves necessarily carried back to the time when the schistus on which we stood was yet at the bottom of the sea, and when the sandstone before us was only beginning to be deposited, in the shape of sand or mud."

"The mind seemed to grow giddy," he went on, "by looking so far back into the abyss of time."

Unknown to Hutton and his friends, the strata at Siccar Point offered far more to grow giddy over than they could have fathomed. The red sandstone that topped it all off had formed right in the midst of the rise of plants and mud, the event that led to the eventual evolution of trees that had made possible their boat and their discovery of deep time, and so many other aspects of their lives.

Our road trip through southwestern Ireland would next take us to an extension of a similar red sandstone on Dingle's northern shores, where it sits dripping with remnants of early roots and mud.

✦ ✦ ✦ ✦

OUR CLOCKWISE JOURNEY AROUND the peninsula had first led us backward in time. But as we rounded the western coast and headed back east along the peninsula's northern shore, we began making our way forward again, drifting out of the Silurian and back into the Devonian when plants and roots and mud were still relatively new to land but were more widespread than they had been at our previous stop.

The guidebook led us to a small parking lot just beyond Wine Strand, a sandy beach named for the petite burgundy cliffs that rise up around it—cliffs made of a similar red sandstone as Hutton had found at Siccar Point.

Cows munched on pasture above the cove that we descended into, where we found a large patch of pale root halos that were similar to the one we had discovered at our earlier stop. But whereas that earlier halo had been small and elusive, these were impossible to miss. Along both edges of the cove, receding into the turquoise tide,

MUD

171

the maroon rock sat tie-dyed in extensive knots and clumps of light gray, downward-facing tendrils. They were everywhere, hanging lung-like and as large as my torso, with thumb-thick pipes narrowing downward to toothpick-thick squiggles at their bottoms. With our feet in the sand, we were standing in the midst of a Devonian floodplain, vegetated with small shrubs that probably grew no taller than our knees but that had roots that reached deeper underground.

Neil called me over to where he stood. I followed his finger to a spot where a thin black line, no longer than my pinky, sat vertically within the middle of a halo.

That, he said, was the remnant of an actual root. Not the chemical trace of a root, but the root itself that had trapped mud and pumped water and nutrients up through stems some 400 million years ago, when the planet was covered in nothing but green stubble.

My mind, to borrow Playfair's phrase, seemed to grow giddy as I touched that moment of growth. How far the Earth system had come since then.

On the opposite side of the cove, Yorick stood smacking his hammer against the rooty rocks to gather more mud samples to bring back to Will. I wandered over to join him and found that the portion of the cliff above the red sandstone contained a layer of darker gray rock full of sands and pebbles of all different sizes. It was a conglomerate, a disorganized dump of sediment typical of a flash flood or landslide.

"An hour's worth of chaos," Neil said.

Based on very loose approximations, he suspected that the red, rooty rock contained perhaps one thousand years of Devonian muds and sands, with only a couple of years' worth of plant growth. The arrival of that dump of sediment on top would have smothered and shaved off a portion of that red rock and roots, leaving behind the thick conglomerate that rose above us.

On top of that "hour's worth of chaos" stood those cows that stared blankly down at us as they chewed. Some were maroon, others were black, and together they matched the colors of the sandstone and conglomerate, almost uncannily so. They looked as if they had been painted by the ancient floodwaters that made the stone that generated the plants they were now eating up.

We all stepped back to watch them for a moment, the tide splashing the root halos below them.

"It's actually quite poetic," Yorick said, the strata-colored cows nodding their heads as they chewed.

Twelve

AS FAR BACK AS I CAN REMEMBER HAVING MUSCLES strong enough to climb trees, I sought refuge in the limbs of a small maple tree that sat in front of the house I grew up in. I would hoist myself up on the lowest branch a few feet off the ground and then scramble to another branch higher up where I would sit and rest. I watched cars driving by and neighbors walking dogs. Mostly, though, I was creating a world in my own head. I pretended that one of the branches was my bedroom. I stashed acorns for squirrels in some of the crooks between limbs and swung from the thickest boughs, doing pull-ups until I grew tired, and then clambered back into the crook that was my bedroom.

That tree was one of the first places outside of my home that felt like home.

An Earth without roots and stems isn't just an Earth with less mud and less oxygen and a less complex carbon cycle. It's a planet without life strong enough to hold us, to ground us in that holding and to expect nothing back from it.

I didn't know this as a kid, but one of the world's oldest fossilized forests sits just ten minutes off the highway I drove each year to summer camp in the Adirondacks. Located in an abandoned quarry just outside the town of Cairo, New York, this collection of trees dates back some 385 million years. Neil, Will, and another colleague from the UK reported evidence of a slightly older forest in South West Wales in 2024, in rocks that date as far back as 390 million years—the oldest known of on the planet.

By that time, plants had evolved more complex features that offered ever more stabilizing material for mud to cling to. Leaf litter gathered in newly forming logjams made of branches that slowed river currents and allowed yet more clays and silts to pile up.

It's tempting to imagine that all of these new plant- and mud-borne developments were strictly positive for the natural world. A confetti of growth that led to an abundance of life forever thereafter. But the reality was far more complicated. Yes, the rise of plants and mud brought with them many benefits. But their global spread also sent a shock to the system that had already existed for billions of years without a stem in sight. The outpouring of nutrients that roots released into waterways grew so vigorous at times that massive marine algae blooms proliferated and created oxygen-depleted cesspools. Some interpretations of the strata from the Devonian suggest that such blooms may have contributed to extinctions that eliminated up to 80 percent of all shallow marine life at that time.

Over the course of millions of years, the system eventually regained balance, and life learned to coexist with the newly vegetated way of things. Early land dwellers, including millipedes and wormlike creatures, evolved anatomies to navigate through the terrestrial muck that they encountered.

"Mud is providing a totally different medium for things to live in," says Anthony Shillito, a former student of Neil's who now works at the University of Saskatchewan.

Whereas an animal burrowing through coarse material, like sand, can move by bulldozing their way along, an animal navigating through mud must contend with the inconvenience of stickiness. A creature like a worm overcomes this hurdle by contracting its body and squeezing water out of its way before flexing out through the gap it has created and moving forward. These movements, in turn, could have helped shape the landscapes of mud—as Charles Darwin astutely observed in his treatise on this species.

Some invertebrates also ingest muds to extract nutrition, then

MUD   175

defecate those sediments back out in slightly altered forms, loosening up and adding to the reservoir of mud as they go. This loosening of silts and clays, in turn, would have helped those sediments drift up and out of rivers and spread inland during floods, making for more expansive floodplains that generated more plants and allowed more invertebrates to travel farther inland.

"They are phenomenal engineers, today and in Earth's past," says Lidya Tarhan, a paleobiologist at Yale University who has studied the biogeomorphology that comes from species like worms.

The extent to which these earliest of burrowers shaped Silurian and Devonian landscapes largely remains a mystery. Their burrows have, for the most part, been erased. We may never know how integral they were to distributing the mud that created the floodplains that laid the foundations of all modern ecosystems we inhabit today.

Neil addressed this conundrum in a 2020 paper that he coauthored on the coevolution of life and strata: "The landscapes and seascapes of Earth's surface provide the theatre for life," he wrote, "but to what extent did the actors build the stage?"

Continued work on the rise of mud will help answer this question.

✦ ✦ ✦ ✦

WHEN I CONNECTED WITH Will McMahon over video chat to discuss this transition to a greener, muddier world and to check in on the results of his microscope work, he emphasized that there was still a lot we don't know—and may never know—about how and why mud accumulated the way it did during this time. It's a slow and expensive process to analyze all of his mudstones from around the world, and not without its hiccups. But from his preliminary findings, he thought he could begin to color in a few more details regarding the extent to which plants trapped mud versus created new mud on land.

It was clear, from the onset of our conversation, that Will shared

Neil's perspective on the seemingly mundane. "People view mudstones as the simplest, most boring rock type," he told me from his apartment in Cambridge. "But they are wrong."

While any given silky gray mudstone may look rather plain to the naked eye, he told me, each harbors a world unto itself on the microscopic level, with an assortment of different textures and histories that ever-improving microscopy has helped reveal in greater detail.

He shared his computer screen with me over our Zoom chat and hovered his cursor above two images to illustrate this point. Both were microscopic snapshots of mudstone samples that were falsely colored with blue, green, and gray specks to represent different mineral types. The shapes smooshed together like a Magic Eye image.

The square on the left, Will told me as he dragged his cursor over to it, illustrated a classic example of a mud with clays that had been physically eroded and retained by plants rather than chemically weathered by them. "In general," he said, outlining the turquoise blob with his cursor, "the physically eroded stuff has very defined edges."

Since these physically eroded clays formed while smacking and pummeling downstream in some sort of channel, they also often display a fluid organization, like the stacked pebbles that Neil had pointed out to me at Slea Head. With Will's assistance, the Magic Eye image came into focus and I could see how the longer, straighter sides of the physically eroded clays had stacked and aligned like books slumped on a bookshelf.

Then he directed me to the square on the right, the chemically weathered clays that had been created by plants rather than just retained by them. Those particles had none of that fluid organization to them, nor did they have the same brittle edges. Instead, they looked like a random assortment of lint.

"They are manic, going everywhere," Will said, swinging his cursor around some of the fuzzy edges.

MUD 177

Instead of organizing fluidly, these chemically produced muds squeeze in among surrounding grains and take on different geometries dictated by the contours of the other nearby minerals.

Will hovered his cursor back over the left square—the physically weathered material—to make another point. "That's all feldspar," he told me, pointing to falsely colored turquoise blobs.

Feldspars are a capricious class of minerals that morph into other types of minerals in the presence of acids that plants release from their root tips. He finds lots of feldspar in the oldest of his samples, from before the rise of plants and mud, but hardly any of it in the younger mud deposits from times when plants were more widespread.

He pulled his cursor back over the right square. There, the feldspars were all but absent, another clue that this sample had been chemically weathered.

Across all the mudstones he has looked at so far, only a negligible amount of manic, fluffy clay particles appear in samples from before the rise of plants, and at least a little bit of it appears in virtually all of his samples from after the rise of plants. Feldspars, which are abundant in earlier samples, are largely absent in later ones.

"And we are looking hard in both," he told me. "It's not like we are putting wool in our eyes to fit a nice story."

Though he's still too early in the process to make any definitive conclusions, these combined observations suggest that chemical weathering did, indeed, increase significantly with the rise of plants on land—and, in so doing, expanded Earth's carbon dioxide sinks. The plant-generated pulse of chemical weathering may have played a role in the global cooling that appears to have chilled Earth through parts of the Devonian.

To have kept Earth from spinning into a runaway cooling loop like Snowball Earth, though, other pieces of the Earth system would have needed to adjust to compensate for the cooling effects of the widening carbon sinks that roots and stems and mud brought with them.

Luckily, the Earth system has built-in response mechanisms to do just this. When the climate cools, the atmosphere becomes drier. Less moisture in the atmosphere leads to less precipitation, and less precipitation leads to less plant growth that, in turn, leads to less chemical weathering. This shrinks the carbon dioxide sink that is chemical weathering and allows the greenhouse gas to pool back up.

By studying how the dance between carbon sources and sinks played out during the rise of mud, we can better understand how those sources and sinks will respond to our own perturbations of the carbon cycle now and in the future, says Michael D'Antonio, a geobiologist at the Field Museum of Natural History in Chicago, who uses computer modeling to study how the rise of plants may have impacted Earth's carbon cycle. "It's a natural laboratory written in the rock record," he says.

The ins and outs of the carbon cycle come down to the conservation of mass. That is, the amount of carbon on Earth has never changed since the planet first formed. This pale blue dot is a closed system. Every particle of carbon here today has been here since the beginning; it has only changed form and location. From rock to algae to bone to blood and back to rock. Without this closed carbon cycle, Earth wouldn't be habitable for us. It would either free-fall into a runaway cooling loop even more intense than Snowball Earth and wind up like Mars, or free-fall into a runaway warming loop and end up like Venus.

Even as human populations have ballooned and carbon dioxide levels have increased in the atmosphere in recent centuries, the total amount of carbon on the planet has remained the same—it has just changed location, from rock to air. Every piece of C that is in your blood, your hair, your teeth and bones, has been here on the planet since the planet formed. We are all recycled pieces of that event.

But while the total amount of carbon on Earth has never changed, the networks and pathways it travels through have changed and gained complexity through time. Understanding how plants and mud added to that complexity is key to making sense of the modern carbon cycle, and the extent to which we have perturbed it.

MUD 179

Based on computer models that D'Antonio and his colleagues have constructed, he thinks that the rise of mud on land was predominantly a product of increased retention of mud rather than creation of new mud. If orders of magnitude more mud had been created anew with the rise of plants, as Will and Neil argue, D'Antonio and his colleagues think that this pulse of chemical weathering would have led to a more extreme instance of global cooling than we find evidence for in the rock record. "Even very small imbalances, when integrated over long time periods, will lead to runaway effects, either to Mars-like conditions or Venus-like conditions," he told me.

Still, he concedes that plants must have introduced some pulse of cooling to the carbon cycle through increased chemical weathering. "It's hard to imagine that not happening, given what we know that plants do," he says.

Neil stands by his interpretations. He thinks computer models that conflict with observations in the rock record are likely missing some intricacy of the Earth system. "If those rocks have been around for 400 million years, and models have been around for 40, it's obvious which one is going to be wrong," he told me.

While their opinions may differ, D'Antonio sees great value in Neil and Will's efforts to continue quantifying and understanding the rise of mud on land.

"It was an absolutely heroic amount of work that they did," he says.

Together, their combined work will help tighten our understanding of the carbon cycle, which will in turn tighten the computer models that are shaping our understanding of how we should behave to help stabilize our climate today.

✦ ✦ ✦ ✦

WITH THE RISE OF computational power over the past couple of decades, more geologists are relying on computer models to draw meaning from the rock record rather than going out to make new observations of their own. They do so, Neil says, because they think that there's only so much more we can gain by continuing to scour Earth's surface; that the strata of the planet offer, at best, a tattered manuscript of the planet's past, as Darwin lamented in *On the Origin of Species*; and that mathematical models can do a better job of interpreting the information we already have and packaging it into buckets of data we can use to assess the present and future of our planet.

With this trend toward computer-based geology, Neil's intense focus on ground-based work has become something of a rarity in the field. But he digs in his heels and insists on conducting the vast majority of his research out in the actual world, with his eyes on actual rock.

While it's true that the depth of knowledge we can gain from strata is inherently limited, Neil sees what is available to us as invaluable grains of completeness in a naturally incomplete record. Some sedimentary layers have certainly become obscured through time and might not offer up reliable information, but others are as well preserved as they possibly can be and offer clear glimpses of what the planet looked like when they formed. A ripple or a raindrop or a clump of mud looked the same hundreds of millions of years ago as they do today. It's these legible sentences in the strata that add so much to our understanding of how the planet has evolved through time and will continue to evolve in the future.

And there are plenty of legible sentences yet to be found, Neil says. His happenstance discovery of the world's largest millipede fossil lying out in plain sight on a highly trafficked beach in England is proof enough of this.

The value of any given sentence within the manuscript of Earth's past can only be measured by how the reader of that manuscript chooses to view the book, Neil argues in a paper he coauthored in 2021. If you're looking to read a novel in the stratigraphic record, then you're going to be deeply disappointed by all the missing pieces and incomplete plotlines. If you're looking to read a series of short stories, you're going to be more pleased. And finally, if you can be contented enough with accepting the stratigraphic record as

MUD

181

something more transactional, like a phonebook, then you will be fully satisfied with what you find. The list of names and numbers—that is, geologic events and their ages—is enough to get the ball rolling on our understanding of the narrative held within the rocks, and then it's up to us to connect the dots and fill in the gaps within the story. An incomplete list, but of legible fragments.

If the Great Oxygenation Event or Snowball Earth or the rise of mud on land are any indication, these listings—both individually and collectively—have plenty to teach us.

"As eerily familiar snapshots from deep history," Neil writes, "they cannot but ground an observer's understanding of their own personal time and place on a long-lived and restless planet."

✦   ✦   ✦   ✦

AS WE DEPARTED THE Dingle Peninsula, I asked Neil how he responds to colleagues who question the value of putting so much time and energy and money into teasing apart ever finer details of Earth's story. The ones who question the point of it all, this work of on-the-ground stratigraphy.

He didn't mince his words.

"One of the silliest questions one can ask is what's the point," he told me. There's really not much point in lots of things, he added. "What's the point in this book? What's the point in geology?"

The jewel-toned sea sprawled out to our right and I had a flash of gratitude, of incredulity that I was sitting there in that car with Neil and Yorick. Gratitude for the collective work that has gone into getting us to where we are in our understanding of Earth history today, to the point that my skin can prickle knowing that we were driving over land that formed when the first plants were just beginning to cling here. There's magic in the unknown, but there's also magic in the known, in knowing all that had transpired to pull those plants into existence and then, from there, to throw us in that car together on that scenic road hundreds of millions of years later, talking about them. How the evolution of bacteria capable of turning sunshine into sugar and a few global ice ages, together, made possible the creation of life with cells complex enough to build stems and roots and leaves and eventually brains large enough to make sense of all this.

"You ask it of an individual thing and there's no point," Neil said with a look through the rearview mirror. "But it's part of the whole. The whole has a point."

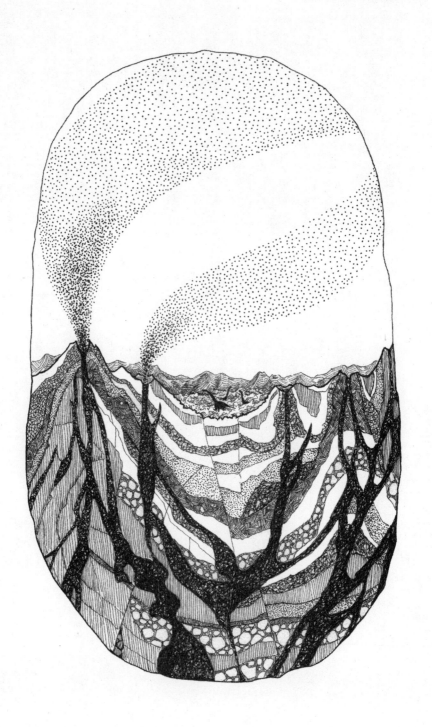

PART IV
HEAT

Light streamed down from the sun, just as it had always done. Now on fern fronds, now on shark fins. Now on an Earth that looked more like our home.

Light arrived as light, but then changed from something seen to something felt. As heat, it radiated and pooled up in webs of carbons bound to two oxygens, those dioxides that erupt from Earth's insides.

Warmth quivered and billowed, caught up in magmatic breaths. This was a good thing, until it wasn't. When breaths snapped to screams and rushed out too fast, heat overflowed and ice bled out.

Lungs withered, oceans darkened, rain soured.

With heat came storms, with storms came lightning, with lightning came fire. And now there was something to burn.

Thirteen

THE HEAT OF THE PLANET PULSES THROUGH THE heart of the oceans in arteries driven by sinking salt. While our own hearts beat with an electric pulse, oceans beat with a pulse propelled by the formation of sea ice at the poles.

When seawater freezes, it leaves behind a trail of heavy, dense saltwater. This cold, salty brine descends as a rapid submarine current that enters the planet's bloodstream in today's North Atlantic and the Southern Ocean around Antarctica. These sites of deepwater formation propel submarine rivers that ribbon and surge alongside continents and in and out of submarine canyons, until they eventually warm enough to rise back up to the surface, where they loop back poleward to begin the cycle anew.

Our blood circulates through our bodies once every twenty seconds. Seawater circulates through the global conveyor belt only once every 1,000 years. Much of the heat that the oceans absorb today might not resurface until the early 3000s, emerging amid our descendants as a hot, nasty reminder of our disregard.

We can't know with certainty what this future will be like for our descendants. But we can get some sense of it by looking at strata laid down during another period of warmth in the planet's past.

If you reach out your arms and imagine Earth's 4.54-billion-year history as a timeline that extends from the tips of your right hand to the tips of your left, we have now reached the middle knuckle of your left middle finger. The

Mesozoic. The "middle life" era (meso-, middle; zoic-, life) that spanned between 252 million and 66 million years ago. It was preceded by the "ancient life" era of the Paleozoic (which began with the Cambrian explosion of life), and was followed by the "recent life" era of the Cenozoic that runs up to the present day. As a reminder, eras are the second-largest of the nested dolls that make up the geologic timeline. Eons, like the Archean, are larger; and periods, like the Cambrian and Devonian and Silurian and so forth, are smaller.

The pathway into the Mesozoic was one of great growth. After mud and plants arrived on land toward the middle of the Paleozoic, stems grew ever taller and animals followed suit, energized by the air that came from the plants. Oxygen levels surpassed today's concentration of 21 percent and peaked possibly as high as 35 percent during the Carboniferous period, which spanned between 359 million and 299 million years ago. With this deep well of oxidative energy, giant dragonfly-like insects swelled larger than seagulls, scorpions lurched longer than woodchucks, and millipedes like the one that Neil Davies found on the coast of England scuttled larger than adult humans. Meanwhile, Earth's landmasses inched together into a single massive supercontinent, Pangea.

This time of expansion may have felt like a pretty good one for the life living through it. And, by many measures, it was. Metamorphosing insects and reptiles emerged for the first time. Plant roots were deepening, new seeds were spreading, and the continents were finding new ways to fit together.

But all good things must come to an end, and this was no exception. The shift in tenor began first as rumblings beneath the planet's crust, imperceptible to the dragonflies droning through ferns and scorpions clicking claws behind dusty boulders. Eventually, those rumblings exploded as a toxic haze of volcanic ash and lava that spread across what is now northern Siberia. Up went the magma through subterranean pipes and out of craters, singeing life around it.

Even this, though, wasn't out of the ordinary—at least, not at first. Life had existed alongside magmas ever since the first cells emerged. A squirt of lava here, an explosion there. It comes with our planet's territory.

But what followed over the next tens of thousands of years proved far more extreme than anything life had contended with before. Across some 5 million square kilometers, Earth's innards emerged just below and atop the surface of the crust, producing enough lava to cover the entirety of the modern contiguous United States with a layer one kilometer thick. The gases burped up with Earth's guts warmed and poisoned the atmosphere and ocean from pole to pole. Surface waters heated up by some 10 degrees Celsius and sea ice vanished. The darkened oceans absorbed more warmth from the sun in a feedback loop that ran in reverse of the ones that had led to the Snowball Earth glaciations. An estimated 96 percent of all marine species and 70 percent of terrestrial species perished through this Great Dying, the single worst mass extinction in Earth history. This time of loss took place some 252 million years ago; its end marks the beginning of the Triassic, the first of the three periods that make up the Mesozoic Era.

Over time, ferns and cattails reemerged through a haze of charcoaled forests and wetlands. Sharks and horseshoe crabs continued pulsing through seas, and dinosaurs arose and began deepening their toeholds on land. But then, just 50 million or so years later, it happened again. Something inside the earth snapped and sent up some 10 million square kilometers of magma, flowing through the seams of Pangea and shaping what would become the bottom of the Atlantic Ocean.

Again, Earth's insides pooled up and out with warming gases; carbon dioxide trapped heat like wool blankets on an already muggy night. Those magmas arose through a series of four pulses that spanned some 600,000 years, annihilating an estimated 76 percent of marine and terrestrial species and marking another one of the Big Five mass extinctions.

"Every pulse of volcanic activity would have acted like an

adrenaline shot to the global atmosphere, shocking the system," writes a team of scientists based in Dublin and Maine who have studied remains of this extinction in East Greenland.

When these eruptions finally slowed, maybe there was sizzle and a sigh. Maybe there were wary reptilian eyes squinting up at a newly burnt horizon. We can't know for sure what this moment looked or felt like. What we do know is that feet hit the ground and stepped forward into a changed world, again.

"There is always life," those East Greenland researchers write, "even during the most extreme of environmental conditions."

Just because there is life, though, does not mean that life exists in robust, thriving communities. The scaffolding that connects living things is more precarious than the living things themselves and takes a long time to rebuild. In the case of the end-Triassic extinction, the diversity of plant communities took hundreds of thousands of years to recover while animal communities took more on the order of millions of years—around 2,500 human generations.

Plenty of other periods of intense volcanic activity have unfolded over the past 4.54 billion years without precipitating anywhere near the same level of destruction. What made these two events different?

I hate to break it to you, but it's this: in both cases, the magma that welled up from the mantle sat directly beneath massive reservoirs of oil, gas, and coal. As that magma rose to Earth's surface, it burned and combusted those fossil fuels, releasing not only carbon dioxide but also toxic butanes and benzenes and ozone-depleting gases.

Perhaps this is all starting to sound familiar to you.

To make matters worse, we're warming the planet through our combustion of oil, gas, and coal somewhere between ten to one hundred times faster than those ancient magmas did. We can look to the Mesozoic not only as a window into the past but as a mirror that reflects what might come of our future, so long as we remember that rates of change matter and that the rates of change taking place today far exceed those of the Mesozoic. Take a glass lid from a freezer and

HEAT

191

place it on a boiling pot of water, and it will shatter. So too will parts of the Earth system if heated up too quickly.

Magmas from the second of those two Mesozoic extinctions sprawl across parts of Africa, Europe, South America, and North America—not in far-flung corners, but visible in plain sight. I drove by some with regularity as a kid growing up in the suburbs of New York City, though I didn't know what they were at that time. I knew them as the Palisades, those burnt umber cliffs that rise out of the western shore of the Hudson River.

I looked at those stony pillars with my jaw slackened and eyes wide from the back seat of our family minivan, but didn't think much of them beyond their beauty. I didn't know that they stood there like an omen to the scores of humans traveling to and from the hub of the world's largest economy, offering a glimpse of what that economy was sowing with its consumption and what our descendants will reap as a result.

✦ ✦ ✦ ✦

THE MESOZOIC OFFERS A chance to witness the power of heat in the Earth system. Unlike the Great Oxygenation Event or the Snowball Earth events or the rise of mud and plants on land, the rise and fall of warmth is not something that has happened once or a few times in Earth history. It is something that has continually happened in the background as all these other changes unfolded. It is the physical manifestation of the carbon cycle, which itself is a product of the entire Earth system at work. It both shapes and is shaped by changes in the atmosphere, continents, oceans, life, and ice. It is not only a symptom of change within this system but also a powerful agent of change, driven by and driving the abundance of carbon dioxide in the air.

Though we've grown wary of such greenhouse gases today, it's worth remembering that we need a certain amount of these molecules

192   STRATA

in our atmosphere to keep our planet habitable. Without carbon dioxide, most of our planet's warmth would slip back toward the sun where it came from. Oxygen and nitrogen gases—the materials that make up some 99 percent of our modern atmosphere—have no insulating powers. Their molecular geometries rest too tightly bound to admit heat, so it streams right past them when inbound from the sun and bounces back to outer space. Molecules of carbon dioxide, on the other hand, have a larger and looser geometry that can trap warmth and keep it here.

The problems, of course, arise when these greenhouse gases pool up too quickly and the heat piles on too fast. I've chosen the Mesozoic to illustrate these effects not because it was the first or only moment of intense heat, but because it provides a broad swath of time to pull insights from. It is an era with some deeply familiar qualities, along with some wholly unfamiliar ones. Only after it ended did placental mammals begin to rise across the continents and, with them, the real possibility of us. The Mesozoic marks the last chapter when our oldest recognizable ancestors were entirely out of the picture.

I equate the Mesozoic with heat because it was overall a much warmer time on this planet. Some climate models suggest that global surface temperatures reached an average of between 14 to 25 degrees Celsius warmer than the modern average, making for little to no ice at the poles and elevated sea levels that would have sloshed the top of the Washington Monument. This era spanned some 158 million years, though, and there was bound to be variability throughout it. Cold snaps and heat waves, periods of freezing and periods of melting. There's much we don't know about the fluctuations during this time, but we do know that on the whole, it was hotter than today and reflective of a version of the world we may be heading toward.

Compared to rocks from the Great Oxygenation Event or the Snowball Earth events, Mesozoic deposits are far younger and so have had far less time to get mangled and melted by plate tectonics.

They are, as a result, more plentiful and easier to read. The grab bag of potential questions we can ask of them is larger, as is the pool of researchers studying them. Some are investigating how marine life responded to warming waters, others are looking at how the distribution of life on the continents changed, and still others are exploring how nonliving entities, like wildfires, responded to the heat.

Rather than choose a single line of inquiry to follow here, I'm interested in drawing from the entire palette of warmth to paint a broad-strokes portrait of this hotter version of Earth. To see what it can reveal about the direction we're heading today, to understand how our pace of change now compares to that of the past, and to explore how we can read any of this in the ground beneath our feet.

Ironically, the strata that provide context for today's warming are some of the very rocks that we have warmed our world with. About 70 percent of all modern oil deposits rest in Mesozoic rocks. They are the fossilized remains of marine plankton blooms that died, sank, and became buried within seafloor sediments that are now extracted and refined as fossil fuels. As we go about our lives and burn those seafloor deposits, we inhale bits of those ancient blooms. We pull them into our lungs and blood, incorporate them into our bones. We send them back into the atmosphere as we carry on combusting, returning the planet back to the conditions that those blooms once thrived within. Maybe this pleases the tiny ghosts of plankton past.

By this point in the twenty-first century, we've all become familiar with the repercussions of excess greenhouse gas emissions. As I write this line in July 2023, the sun outside my window in southern Maine hangs heavy and orange from smoke blowing west from the most destructive wildfires ever to hit Nova Scotia. The skies of New York City have taken on an even deeper acrid tint with smoke funneling south from fires in Quebec. Manhattan joined Chicago last month with the lowest-ranking air quality of all major cities worldwide, the result of unprecedented Canadian fires that will lead to some 200,000 evacuations, including the displacement of tens of thousands of Indigenous people. Meanwhile, precipitation records will shatter across parts of New England with rains that will destroy more than 2,000 acres of crops and result in upward of $15 million in losses in Massachusetts alone.

Largest fires. Worst air quality. Most moisture. We've reached an era of superlatives.

Here on Maine's coast, I sit at an epicenter of another superlative: fastest warming seawater. The combination of the Gulf of Maine's bathtub shape and its location at the confluence of two rapidly changing currents has resulted in a warming that has outpaced 99 percent of the rest of the ocean. The top five hottest summers on record for this body of water have all taken place since 2012. The fabric of the marine ecosystem is rapidly shifting, along with the livelihoods of the humans who depend on that ecosystem. Shrimp and cod populations have declined, and phytoplankton that make up the base of the food chain have grown less productive. Meanwhile, longfin squid, black sea bass, and other warmer water species have begun traveling up here with increasing regularity. In 2018, a lobsterman working off the coast of Boothbay pulled two seahorses from his traps—a species more typically found between South Carolina and Cape Cod.

I feel the eerie warmth on my own skin as I dunk in waters that should give me brain freeze but instead feel as comfortable as bathwater on the hottest of days. These warming seas not only risk losing their native species; they also struggle to hold on to their carbon dioxide. Warm liquids can't trap dissolved gases as well as colder liquids can. You've experienced this if you've ever had a disappointingly flat sip of a warm carbonated beverage.

As oceans warmed during the Mesozoic, they too would have released more carbon dioxide into the already warm atmosphere, in a self-perpetuating loop of heat that would generate more fires and more storms and poorer air quality. Heat that would, yet again, force the Earth system to find its way back to balance.

HEAT

195

✦ ✦ ✦ ✦

ON A THURSDAY MORNING in mid-February, I drove north on I-95 to Waterville, Maine, to meet with Ian Glasspool, a Colby College paleobotanist who researches the planet's oldest evidence of wildfire. He was part of the team of researchers that studied the end-Triassic extinction in strata from East Greenland. There, they found a striking fivefold increase in charcoal across that period of warming—evidence of a dramatic uptick in blazes at that time. Those charcoal remains painted a picture of a world on fire. I traveled to Waterville to meet with Glasspool and learn why, exactly, increased volcanic gases led to this massive spike in wildfires, how we can read this in the rock record, and what this might tell us about wildfire patterns of the future.

As my tires hit pavement, melting snow steamed off the road in a haunting mist. That day would break records in New England as one of the warmest in February, reaching beyond 60 degrees Fahrenheit when temperatures should have hovered closer to freezing. Air that should have stung cold and dry against my cheeks streamed soft and humid through my car's open windows.

I met Glasspool in his office on the fourth floor of a brick building in the center of a campus where students walked in sandals alongside piles of mushy snow. We sat down together at a table where a collection of small resin discs rested on a piece of cardboard. "They are tedious-looking things," he said, picking one up, "but there's a whole heap you can get out of them."

He spoke in a British accent that had been mangled by a couple of decades spent here in the States. With the disc in his hand, he lifted his wire-rim glasses up over his head of salt-and-pepper hair and pointed to the bits of charcoal hovering within it. They were no larger than the tip of a pin, nothing I would think twice about if he hadn't pointed them out.

From the mouth of a small jar, he dabbed a drop of oil onto the disc, then placed the specimen under the lens of the microscope in front of him. Once he pulled it into focus, he hooked the microscope up to his laptop and pointed to his screen. "See how it's silky there?" he said, jiggling his cursor over what looked like a magnified fleck of lint, not a window to the ancient atmosphere. "That's charcoal."

The charcoal was not to be confused with coal, Glasspool explained, which gradually darkens from brown to black as it is buried, compressed, and heated within the earth. Whereas charcoal has a silky luster, coal appears more matte. He shuffled around on his computer and brought up another image to show me the difference. "That's coal," he said, hovering his cursor over a black speck with less luster.

I was delighted to discover this niche area of research, and humans such as Glasspool who could not only tell the difference between these lint-sized black specks but also pull heaps of meaning from them. He's now a leading figure in the field of ancient wildfire studies, but he didn't begin his career in the geosciences. He first studied biochemistry and worked as a sales rep for a biomedical company. But at the age of twenty-five, he found himself rather unhappy in that job and decided to return to college to study engineering geology. On a whim, he sat in on a paleontology lecture on microfossils. That particular day, his professor was offering up a can of baked beans as a prize for the best answer to one of his questions.

To Glasspool's surprise, he won.

"At the time, a can of baked beans was quite tempting," he told me. "One less day of shopping. Fantastic student diet."

With his interest piqued, he arranged to audit the course for the rest of the semester. Soon, he changed tracks entirely to pursue a degree focused on paleontology. Now he is credited with identifying the oldest charcoal in the world.

More precisely, Glasspool has identified the world's oldest charcoal on *two* occasions in his career. First in 2004 when he found 420-million-year-old deposits in rocks from the Welsh Borders, and then again in 2022 when he pushed the date back another 10 million years with a different set of Welsh samples.

HEAT   197

"Half of what makes geology so fun is you never know what mystery is going to turn up," he told me.

To study remnants of these ancient fires, he first dissolves his rock samples in acid, then sieves out the tiny black specks left behind. To manipulate and orient each fleck for microscope analyses, he uses a wooden skewer that has a single whisker from his cat, Bingo, duct-taped to the end.

"Low budget, do-it-yourself," he told me. If he used a store-bought paintbrush, his tiny samples might get caught up in the hairs; Bingo's whisker lends him more control.

Viewed with a simple light microscope, these charcoals reveal the intricate contours of cellular walls that have remained pristinely preserved for hundreds of millions of years.

The very oldest charcoals that Glasspool has studied come from the Silurian period, when vegetation was only just making its way onto land and was still quite diminutive. Those low-to-the-ground relatives of mosses and liverworts were not only petite but also moisture-loving. As we spoke about them, it was hard for me to imagine those moist growths flaming hot enough to fuel the earliest wildfires.

Glasspool and his colleagues agreed. He clicked around on his computer and pulled up an artistic rendering of an enigmatic growth from a group of organisms that they think may instead have been responsible for producing the earliest fuel. The growths may not have been plants at all; in the fossil record, they often look more fungal than vegetal. Called nematophytes, they grew in tall humps that topped out at more than twenty feet high, perhaps behaving as lightning rods on otherwise barren landscapes.

To my eye, the illustrations of these growths looked like saguaro cactuses I had seen in parts of Arizona, albeit without the spikes and flowers, and I told Glasspool as much.

"What they look like," he responded matter-of-factly, "are giant phalluses." He wasn't wrong.

Whatever the fuel had been, the presence of charcoal that far back in time indicates that enough oxygen had pooled up in the atmosphere to support flames. Fires can't easily burn at oxygen levels below 16 percent, so the presence of charcoal provides a lower limit for oxygen levels during the Silurian.

By the end-Triassic extinction some 230 million years later, oxygen concentrations had grown greater, as had the collection of kindling available to burn on the landscape. Ginkgoes and conifers made up the canopies of forests that grew understories thick with ferns, palm-like cycads, and bennettites—a now-extinct order of seed plants that looked sort of like a palmy shrub.

This highly vegetated hothouse world had plenty of material to start fires and plenty of oxygen to sustain them. The stage was set, and the lights came on with the volcanism and resulting warming gases. But how, exactly, had those greenhouse gases led to the five-fold increase in charcoal that Glasspool and his colleagues had found across that interval?

In some ways, this is a no-brainer. "If you've dried your vegetation more and it's hot," Glasspool said, "it makes combusting a lot easier."

But the heat would have instigated wildfires in less obvious ways as well. Atmospheric warmth increases rates of evaporation and thickens the atmosphere with water vapor that, in turn, leads to more intense thunderstorms. With thunderstorms come lightning strikes, which were the leading source of ignition during the Mesozoic and continue to be the leading natural ignitor of wildfires today. Just 1 degree Celsius of warming in the modern atmosphere may increase the rate of lightning strikes by some 40 percent, according to one study out of the UK.

Beyond more lightning and drier fuel, those Mesozoic eruptions had another effect on the landscape that may have fueled more intense wildfires. In fossilized plants collected from East Greenland, Glasspool's colleagues found that the shape of leaves changed significantly as the end-Triassic extinction progressed. Narrow-leaved species doubled in abundance, while broad-leaved species largely drifted out of the fossil record. This, he and his colleagues reported in a paper on the findings, was likely an adaptive response to increased atmospheric temperatures, since smaller leaves can rid themselves of excess heat more efficiently than larger leaves can.

Just as ripped-up shreds of newspaper burn faster and hotter than intact sheets, so too did these narrower leaves. They made for far more efficient kindling, and they intensified the fires that the uptick in lightning strikes had ignited.

These infernos fed the feedback loop of warming by combusting carbon-rich plant material and releasing more greenhouse gases into the atmosphere. But they would have left a more nuanced legacy within the carbon cycle than just this. Their smoke could have reached all the way to the stratosphere, where it would have blocked out warmth from the sun and produced a global cooling effect. By destroying large expanses of vegetation, the fires would have also increased rates of erosion in some places, burying dead organic material that would have otherwise decomposed and released carbon dioxide into the atmosphere. In this roundabout way, those Mesozoic wildfires could have served both as a source and as a sink of carbon—as they continue to today. We know all of this not so much from the rock record but from modern studies of fires.

With increased erosion, nutrients like phosphorus could have spilled into the ocean in larger volumes than they had before. This may, in turn, have increased algal blooms that would have increased oxygen levels that would, in turn, have increased the possibility of fire.

These ancient insights provide a clearer sense of the global phenomenon that is fire, Glasspool says. But they alone can't help us predict wildfire patterns on a local or regional level today. Human behavior and land use practices are just as important to consider as climate dynamics when predicting and managing modern burns. Still, insights from the strata can demonstrate how fire plays into the feedback loops that shape global climate over geologic time and this, in turn, can help climate modelers make more accurate global projections of the future.

"The geologic record shows that it is a lot more complicated than 'it gets hot, there will be more fires,'" says Jennifer Galloway, a paleoecologist with the Geological Survey of Canada who advocates for studying ancient wildfires as a way to improve our understanding of climate dynamics. Fire, after all, licks through every facet of the Earth system.

"It is accounted for in some of the climate modeling, but I am not sure that its role is appreciated to the extent that it needs to be," Galloway told me.

Fernanda Santos, a staff scientist at the Oak Ridge National Laboratory in Tennessee who studies modern fires in Alaska and works closely with climate modelers, agrees that insights like Glasspool's can improve future projections—particularly since climate models often pull from satellite imagery of fires dating back only about one hundred years. The rock record expands that timeline exponentially.

"I really value ancient data because they can give us this new perspective and new baseline," Dr. Santos said.

The technology needed to study them is also rather simple and affordable compared to other areas of stratigraphy—and, in the case of Glasspool's whiskered skewers, rather cute. If only Bingo understood his role in our understanding of the Earth system.

✦ ✦ ✦ ✦

TO DRAW THE CONNECTION between warmth and wildfires during the end-Triassic extinction, Glasspool and his colleagues didn't just study proxies for wildfires—they also pulled on proxies of paleotemperatures.

A whole suite of paleotemperature proxies exist across strata deposited in different types of environments. In the case of the end-Triassic extinction, Glasspool and his colleagues looked to the plant fossils they gathered in East Greenland, with a focus on microscopic pores called stomata. These tiny holes grow on the undersides of leaves and serve as a plant's portal to the outside world. They pull in carbon dioxide and puff out oxygen and water

HEAT

201

vapor throughout the day, opening and closing with the puckering of cells that look like microscopic lips. Thousands of these little mouths can form within a given leaf depending on the species and their growing conditions.

In their analyses of these fossilized pores from East Greenland, his colleagues found that the density of stomata fell markedly in the time leading up to the end-Triassic extinction. They attribute this striking dip in stomatal density to a possible quadrupling of carbon dioxide levels in the atmosphere—which they estimate led to a 3 to 4 degree Celsius increase in global atmospheric temperatures at that time.

They were able to draw these conclusions from those microscopic plant mouths thanks to more than a century's worth of research before them. Botanist Sophia Eckerson, for example, had studied these tiny pores as a graduate student at Smith College in the early 1900s. "For some educational purposes," she begins her 1908 report of her findings in the *Botanical Gazette*, "it is needful to know which plants possess the largest, the most numerous, the most readily observable, or the most definitely distributed stomata." Little did she know how needful we would become of those observations in the twenty-first century to improve our understanding of global climate.

Eckerson had peeled off the epidermis of more than three dozen leaf species from her college greenhouse, dropped them in jars of alcohol, and arranged them under her brass microscope's eyepiece to get a closer look. Through hours hunched over that lens, she recorded differences in the stomata of a wide variety of plants. She found that figs, beans, and pumpkins had the greatest quantity of stomata, and that wheats, oats, and primroses had the largest individual pores. Most pertinent to today's paleoclimate studies, though, were her observations of the circadian rhythms of the stomata. On most days, they opened to their widest by 10 a.m., began to close by 2:30 p.m., and closed entirely between 5 p.m. and 6 p.m. On particularly hot days, however, they closed as early as noon. "Probably because of incipient wilting of the leaf," Eckerson noted in her 1908 report of her findings.

The leaves, it seemed, pursed their lips to hold on to water in the presence of excess heat.

Fast-forward to 1987, when British botanist Ian Woodward reported that leaves not only deactivated their stomata with heightened temperatures but, over time, lost some entirely. He found a 40 percent decline in the density of stomata in herbarium samples collected from 1787 to 1987—the same interval when carbon dioxide levels increased some 21 percent with the industrialization of the modern world.

Glasspool's colleagues in Ireland have used these observations, along with studies of modern plants, to develop equations that allow them to back-calculate carbon dioxide levels based on the density of stomata in fossilized leaves. It's with these equations that they calculated that the drop in stomata during the end-Triassic extinction reflects a possible quadrupling of carbon dioxide levels during that time. Layer on the fivefold increase in charcoal that they observed across this extinction, and these proxies begin to connect the dots between increased greenhouse gases in the atmosphere, temperature, and wildfire intensity. The proxies can only provide fragments of the whole picture of the end-Triassic but, even so, offer valuable context as we barrel toward a global temperature rise of 2 degrees Celsius by the end of the century—or possibly as soon as 2050. The end-Triassic strata should give us pause.

✦ ✦ ✦ ✦

ACROSS THE ROCK RECORD, other paleotemperature proxies have helped demonstrate how warmth has waxed and waned through geologic time. Some such proxies are fairly straightforward and visible to the naked eye. The glacial dropstones that Brian Harland found in Svalbard strata, for example, indicated that temperatures had dipped below freezing in that location in the deep past. Proxies for warmth can be similarly simple to detect in the rock record. As far back as the 1600s, British physicist Robert Hooke correctly assumed that the turtle fossils that he found on the southwestern coast of England were indicative of earlier warm times there, since turtles like the ones he found don't exist in the modern-day North Atlantic. Ancient remains of alligators and tortoises in the Canadian High Arctic similarly indicate that this region was also much warmer in the past. What's now a dry and frozen tundra once resembled the swampy wetlands of modern-day Florida.

"This is a first order, 'Oh, it must have been really warm,'" says Jessica Tierney, the paleoclimatologist at the University of Arizona who I spoke with about the value of studying Snowball Earth, and who was a lead author of the Intergovernmental Panel on Climate Change's Sixth Assessment Report in 2021.

Tierney is a strong advocate for using stratigraphic data and paleoclimate proxies from hothouse periods such as the Mesozoic to get a better sense of what's to come in our future. While computer models help predict how changes in climate *may* unfold as carbon dioxide levels increase, she says, geologic proxies indicate how the planet *actually did* behave as carbon dioxide levels increased in the past. She is working to help strengthen existing climate models by incorporating new information gleaned from these past snapshots. "Even though we are looking at something in the deep, deep past and the world was somewhat different, the physics of the climate system are the same in both cases," Tierney told me over a video chat from her office in Tucson. "They obey the same rules."

The climate models she is working to fine-tune will help policymakers develop more informed decisions around climate adaptation and mitigation.

Her proxies of choice include fossilized fats left behind from the cell membranes of ancient marine algae. Based on studies of these fats in the modern world, she knows that they tend to contain more double bonds in cold conditions and, as conditions warm, they drop those double bonds. This structural shapeshifting helps the algal cells maintain fluidity as conditions around them fluctuate.

Studies of this shapeshifting in modern algae provide a rubric of sorts that Tierney and her colleagues can use to calibrate their measurements from the fossil record. With these points of comparison, they have been able to estimate that the Arctic wasn't just a little bit warmer when those reptiles lived up there—it may have reached temperatures beyond 100 degrees Fahrenheit.

Geochemical proxies like these can help improve the precision of computer models by providing a known point of comparison from the past. If models can successfully "predict" these known conditions, then scientists can more confidently turn to those models with questions they don't already have answers to.

Beyond these fat thermometers, paleoclimatologists turn to a whole suite of other paleotemperature proxies to fill in our understanding of past climates. They dig into the physical and chemical properties of materials like bones, teeth, and seafloor sediments to gather measurements of ancient temperatures and track past periods of warmth. And—in the case of one research group—they turn to fossilized poop.

The niche nature of this last line of research has less to do with any stigma around poop and more to do with the real challenge of finding fossilized feces in the rock record. This is both because poop doesn't fossilize as readily as something harder like a bone, and because it's often indistinguishable from other rubble.

"There are a lot of fecal-shaped rocks," says Karen Chin, a paleontologist at the University of Colorado Boulder, who has been using these types of fossils to reconstruct ancient environments for more than twenty years.

Chin became familiar with coprolites—the technical term for fossilized feces—early in her career while working as a lab technician for Jack Horner, a renowned paleontologist who inspired the protagonist of *Jurassic Park*. When Chin came across information about coprolites one day on the job, she was surprised to learn that feces actually fossilized, and that people studied those fossils. "I just thought that was the funniest thing," she told me over a video chat from her office in Boulder. She approached Horner to see what he

HEAT

205

knew about coprolites, and he confirmed that he was not only familiar with them but had a sample of one from a dinosaur that he was happy to show her.

To Horner's eye, the bits of chewed-up plant material in that sample reminded him of his days growing up in Montana, when he shoveled elephant dung when the circus came through town. For Chin, the sample brought back memories of working as a trail interpreter in national parks. She would spout off trivia about what the size, shape, and contents of scat can tell us about the animal that left it behind, but not once had she considered that the same could be true of ancient scat. She's now one of the world's leading coprolite experts, has coauthored a children's book on coprolites called *The Clues Are in the Poo*, and is working with researchers around the world to reconstruct a variety of ancient environments—including the hothouse of the Mesozoic.

Compared with other paleoclimate proxies that might provide an average temperature estimate over thousands or millions of years, coprolites uniquely capture a very finite snapshot of just a day or two, or however long the material sat in a digestive tract.

"It's like trolling the waters now and just grabbing a sample," Chin says.

Some of the most informative coprolites she has studied from Mesozoic times come from Devon Island in the Canadian High Arctic, where she traveled with a few colleagues back in 2003 and collected some 400 samples that they are still analyzing today. When their studies are complete, the team will return their samples back to the Government of Nunavut, which granted them permission to conduct their field research.

The team reached their base camp on Devon Island by way of prop plane and helicopter, and traveled by foot up steep terrain to access their collection site along the side of a mountain. They carried guns to protect against polar bears, but found that the landscape itself was the more persistent predator. The thawing summer ground would cave in beneath them as they walked, forcing them to run so that they wouldn't sink too deep and get stuck. They ran over melting ground, crawled through scattered patches of thigh-deep snow, and crossed raging rivers, all while carrying dozens of pounds of ancient poop and other fossils on their backs.

They distinguished the coprolites from the surrounding rubble by their globular, amorphous shapes that resembled modern digestive systems, and sometimes by the presence of undigested bits of food such as squid parts and lobster carapaces. While they ideally would have collected their samples straight from the strata, the steep slopes they were working along had eroded over time and left most of the coprolites in loose heaps at their feet. Chin gathered her samples simply by plucking them up off the ground.

Beyond the coprolites, she and the two other paleontologists on the trip also collected bits of fossilized sponges, sturgeons, sharks, flightless birds, and other remains that helped bring the Mesozoic marine ecosystem into focus. The lone sedimentologist on the trip, meanwhile, focused on collecting mudstones. "We just gave him so much grief," Chin told me. *Oh wow, so fascinating, John*, they would say, ribbing him. *That mudstone really does look interesting, huh.* (Think what Neil Davies would have said!)

With time, though, the paleontologists discovered that the mudstones offered far more than met the naked eye. When placed under a microscope back at their labs in the US, those "boring" rocks revealed thousands of microscopic plankton fossils, indicating that the ancient waters supported expansive algal blooms. "Those sediments were the most interesting things of all," Chin said.

To estimate marine sea surface temperatures from their samples, her colleagues later analyzed fossilized lipids from cell membranes of other ancient marine microorganisms preserved within the sediments and coprolites, using temperature calibrations that Jessica Tierney and others had published. With these proxies, Chin's colleagues estimated that Arctic marine sea surface temperatures rose

HEAT

possibly as high 69 degrees Fahrenheit—comparable to late spring temperatures off the coast of Maryland today.

The physical contents of the coprolites also indicated that the marine food chains of that time were relatively short. Large things were eating very small things. This paradigm continues to hold true in today's Arctic food chain. The players in that food chain were different during the Mesozoic—we don't have large marine reptiles like plesiosaurs—but we do have the same general setup of big things eating small things.

In some ways, Chin says, it's comforting to know that the Arctic can maintain some of its ecological structure even through major climatic changes. But before we take too much solace in this, she reminded me, we must remember that those Mesozoic ecosystems had orders of magnitude more time to adjust to change than modern ecosystems do today.

During the end-Permian extinction that marked the beginning of the Mesozoic, average sea surface temperatures increased by roughly 10 degrees Celsius over a span of some 60,000 years. By the end of the twenty-first century, sea surfaces may have warmed by upward of 4 degrees Celsius over just 200 years.

"There are certainly organisms that are much more adaptable, and those are the kinds of organisms that have lasted a long time despite changes," Chin says. Bacteria will certainly survive through this current period of warmth. Shelly creatures called lingulid brachiopods are also particularly adaptable and will likely persist among other hardy organisms, Chin told me.

But our current breakneck speed of change will test the limits of ecological stability in many other organisms, particularly at the poles where the pace of warming is greatest. These polar species will be among the first to face severe stress from global warming and may have the least time available to undergo adaptive evolutionary change. And unlike species elsewhere, they won't have anywhere to migrate to seek cooler temperatures.

At lower latitudes, thin-skinned amphibians such as toads, frogs, and salamanders have demonstrated more sensitivity to environmental change than any other vertebrates. A 2023 study from the International Union for the Conservation of Nature Species reported that some 40 percent of amphibian species are threatened with extinction—the result not only of warming but also habitat loss and pollution.

Recent evolutionary biology research suggests that species can adapt and evolve on human timescales—a pace far faster than Darwin ever thought possible. But this clip may still not be fast enough. Continued work on paleotemperature proxies and animal fossils across periods of warming will help clarify how communities respond to heat over geologic timescales.

Fourteen

THE NEXT MASS EXTINCTION AFTER THE END-Triassic wouldn't come for some 135 million years, with the asteroid strike that ended the non-avian dinosaurs and marked the close of the Mesozoic. Before that devastation, the intervening Jurassic period was a time of ecological rebuilding. Ferns and ginkgoes rebounded, and relatives of modern-day palms sprawled out to shade the frogs, salamanders, crocodiles, and small mammals who roamed alongside the dinosaurs who were expanding their dominion. The poles were still largely ice-free and, in some places, thickly forested.

This was roughly 100 million years before the birth of the modern Rocky Mountains, when Africa was pulling away from the eastern coast of North America and the Atlantic Ocean was developing as a young sliver between the dismantling pieces of Pangea.

As the creatures of this time ate and nested and hunted in their hothouse world, they would have had no way of knowing that the planet was far warmer than it had been in the distant past. They couldn't have known that the ice that had once covered the poles had largely vanished, that sea levels had risen, that oceans had temporarily acidified and lost much of their dissolved oxygen. They had no way of knowing how the different facets of the Earth system stepped up to ease the stress. How soils and oceans and leaves and so many other pieces of the planet behaved as carbon sinks, absorbing the volcanic gases, pulling them out of the atmosphere and, over geologic time, helping establish a new stability.

210   STRATA

What these creatures probably did know, however, was that dinosaurs were now in charge.

"Every animal larger than a meter in size, for 160 million years, is a dinosaur," says Susie Maidment, a paleontologist at London's Natural History Museum who studies fossils from across the Mesozoic and is the world's leading stegosaur scientist.

Susie thinks that dinosaurs have a lot to teach us about the future of life on our warming planet. This may seem a bit odd—looking to this extinct group of giant reptiles for advice about the future. But dinosaurs held ecological roles very similar to those large-bodied mammals do in today's world, and had roughly the same number of species. Their fossils are relatively easy to collect, and they have been widely gathered around the world for more than 150 years. "They actually make an ideal model organism," Susie says.

She occasionally gets pushback from people who suggest that the science is more flashy than serious, and that it might be a waste of time and money to study dinosaurs just because they are cool and novel. But Susie doesn't see them that way. Cool, sure, but dinosaurs weren't particularly novel during the Jurassic. They were an incredibly diverse group, occupying every ecological niche that exists on land today. Throughout the many millions of years they lived on the planet, the small mammals scurrying around the feet of dinosaurs would have been the novelties. Dinosaurs would have been the given.

"So why wouldn't you study them?" Susie says. "They are the most obvious thing."

I met up with her at a dinosaur dig in the badlands of northern Wyoming to learn more about the Mesozoic hothouse world through this dinosaur-centric lens. Our phone conversations leading up to the trip had been nothing but polite, but in the field, I found that swear words flew out of Susie's mouth as fluidly as her interpretations of the bones. Press officers at the institutions where she has worked have pleaded with her not to use words like *crap* in interviews, but that didn't slow her down as she spoke with me.

HEAT

211

"Obviously I haven't learned what not to say to journalists," she deadpanned as she rearranged her tool belt.

In reality, though, she knows exactly how to speak with journalists, and the expletives simply punctuate her points. She describes her research with more wit and clarity than most scientists I've met, and she knows how to frame a good story.

"We think of dinosaurs as really, really well known," she told me over a phone call in the lead-up to the dig. "But they are actually not that well known at all."

We know *that* dinosaurs lived, thanks to the fossils and footprints they left behind in the rock record. And we have a pretty good idea of what individuals may have looked like, thanks to ever-improving scanning and imaging technology. But beyond these rather focused anatomical details, our broader understanding of dinosaurs as a complex community of creatures falls off pretty steeply. Bones alone can't reveal how one individual related to another, or how they evolved through time in their warmer world. We can only begin to color in these details by gathering a chronology of their lives from the remnants of mudflats, coastlines, and floodplains where they hatched, ate, stomped, and died. That is, from the strata. Trying to make meaning of the bones without considering these layers is like studying a character without any sense of their setting in time or space.

More often than not, though, the strata are not where paleontologists focus their attention. They stick to what they're trained to look at. The fossils.

Susie is bucking that trend within the Morrison Formation, a famously fossil-rich suite of Mesozoic rocks that stretches from New Mexico in the south all the way up through Montana and southern Alberta and Saskatchewan in the north. This expansive collection of sandstones and siltstones stretches across some 1.2 million square kilometers and has spilled out more dinosaur bones than any other rock formation on the continent. *Stegosaurus, Brontosaurus, Diplodocus*—all the characters you learned by name when you were seven years old. Since the 1870s, paleontologists have excavated dozens of these remains, propped them up on museum displays, and captured the imagination of audiences of all ages around the world. They have examined their finds using powerful imaging technologies, published research papers on their discoveries, and have generated millions of dollars in funding to return, time and again, to excavate more fossils.

But all the while, they have had little to no understanding of how one set of bones relates to another in terms of their age. A handful of radiometric dates gathered from zircons (those crystal grains with radioactive uranium that decays to lead at a known rate) have helped constrain the broad package of the formation to between about 156 million and 147 million years old. But beyond these lower and upper bounds, the ages of individual layers within the formation remain largely unknown. As a result, the narrative of some nine million years has been collapsed into a single bucket of time.

That's a big mistake, Susie says. By clumping all the Morrison fossils together, her colleagues have erased the nuances that might reveal how animals evolved through time or shifted geographic ranges in response to the extended period of warmth. And they have also smushed together a whole load of different animals that never would have coexisted. "It just doesn't make sense from the point of view of what we know about evolution," she says. By way of comparison, just twelve million years or so of evolution produced humans, gorillas, and chimps from the same common ancestor.

Susie believes the strata can help us course correct. She's rolling up her sleeves and collecting as much stratigraphic information and datable material across the Morrison as possible, and then breaking it into smaller, more manageable buckets of time that she is arranging into a comprehensive chronology. By sorting and filling those proverbial buckets with the fossils that belong within them, she will generate something like a stop-motion picture of how life evolved over the course of millions of years within a climate that had settled at temperatures warmer than today's.

"She's really someone who is pushing ahead with that in a way that I don't think other people have," says Roger Benson, a paleobiologist at the University of Oxford who also studies life of the Mesozoic. He sees the Morrison as something like a Rosetta Stone for other rocks of the same age, because it is comparatively better studied than almost every other assemblage in the world of that age. "The work she is doing is really important and fundamental," he says.

That work consists largely of collecting stratigraphic logs, centimeter-by-centimeter observations of sedimentary layers from the base of a rock formation to the top such as those that I collected during my time in Australia. As with our Australian logs, the individual barcodes that Susie collects can reveal only so much on their own. But with many compiled together, they illustrate how landscapes transformed over time—in this case, from wet to dry to coastal to riverine, each iteration layered one on top of the next.

Since Susie first visited the Morrison as a graduate student at the University of Cambridge in 2006, she has collected more than 20 stratigraphic logs and has worked to correlate them with some 245 additional ones that others have collected over the years. The end product will be a massive, multidecade effort accomplished by many scientists, but she will be the first to pull it all together into a cohesive package.

The margin of error in studies of Morrison fossils has historically been rather large, says paleobiologist Anjali Goswami, a colleague at the Natural History Museum, London who studies vertebrate fossils from other parts of the world. "She's doing a lot of really time-consuming fieldwork to try to remedy those errors."

The chronology will help answer a whole suite of broad-stroke questions, including how the distribution and diversity of life today compares to the Mesozoic. On the modern Earth, the planet's biological productivity peaks around the humid tropics and then decreases toward the poles. Within the warmth of the Late Jurassic, however, the rock record suggests that equatorial regions grew more arid and less biologically productive than higher latitudes. With a spread of strata that spans some 12 degrees of latitude, the Morrison offers a rare opportunity to look at the distribution of life in these warm conditions across a rather wide expanse.

"It is a perfect case study," Susie says. "We can go out and actually examine some of these really broad questions about how evolution works and how biodiversity is distributed."

Mesozoic strata sit like a giant window across a warming North America, beckoning us to take a closer look.

✦　✦　✦　✦

I ARRANGED TO MEET Susie and her graduate student at the time, Joe Bonsor, on the side of a dusty road in the small town of Lovell, Wyoming, to travel to the dig site together. I stood on a sidewalk with my thumbs dangling from my backpack straps, as earnest as a kid waiting for a school bus. This would be my very first dinosaur dig, and the school-aged child in me was still pinching myself.

Susie and I had only ever spoken over the phone leading up to the trip, so I wasn't quite sure who I should be looking out for as I waited. Her headshot online showed a woman in her thirties with crisply cut blonde bangs, bright blue eyes, a knowing smile, and arms folded across a polka-dotted blouse. I glanced around to look for that face, but all I saw was a mustached man driving by in a green sedan. A block farther down, a man in a baseball cap and a woman in a cowboy hat leaned against the bed of a black pickup truck flicking cigarette butts. I turned around to keep searching for the paleontologists from the UK I was supposed to be meeting, but then heard the woman in the cowboy hat call out to me. "Laura?" she said in a British accent.

This, I realized, was that same woman in the polka-dotted blouse, dressed for the field.

HEAT

215

I jogged over to join them, greeted by Susie's energetic handshake and profuse apology about a snafu the day before. The two quarries they were working out of had very spotty cell service, so we hadn't been able to connect until the afternoon, when it was already too late for me to join them at the dig site. I had spent the rest of the day wandering around the sculpted strata of the badlands close to town, thinking about time and our place in it and what one afternoon really is in the scheme of things. I assured her, genuinely, that it had been a day well spent. We still had several days ahead of us—plenty of time to sink into the Mesozoic together.

The bones she and her colleagues were digging up belonged to sauropods, the long-tailed, long-necked group of dinosaurs that were the largest animals ever to live on land. Each hauled around bodies heavier than two humpback whales, sustained with diets consisting entirely of leaves, stems, and other vegetation. Commercial fossil diggers had found these bones years earlier, but they abandoned them once they discovered how troublesome and expensive they would be to dig up. "They very much trophy hunt," Susie told me. "They want something with big pointy teeth that they can sell quickly for loads of money."

Even so, we had all been sworn to secrecy about the dig's location lest other collectors develop an appetite for a challenge. We were only permitted to take photos on our phones if they were in airplane mode, to avoid geolocation, and we were prohibited from publicly posting any photos with identifiable features on the horizon. When my Airbnb host expressed curiosity about where I was heading that morning, I skirted the question and the follow-up questions. Finally, I had to explain that I wouldn't be able to tell him exactly where I was going.

Paleontologists from the Children's Museum of Indianapolis, who were leading the dig, had secured a twenty-year lease on the land from the rancher who owned it. They named the dig site Jurassic Mile and extended invitations to Susie and her colleagues from London along with others from the University of Manchester and the

Naturalis Biodiversity Center in the Netherlands to join in on what became a large international collaboration dubbed Mission Jurassic. Each team member brought with them their own unique expertise. Some were focused entirely on dinosaurs, others on plants, still others on invertebrates. Together, they were working to dig up material both for museum display and for research purposes that would help bring the Mesozoic hothouse to life.

Susie and Joe were the lone paleontologists who came with a background in geology, and an interest in the strata surrounding the bones. They were happy to help excavate new specimens, but they were even more eager to get out and acquaint themselves with the geology of that portion of the Morrison Formation, which had never been closely studied.

The three of us piled into the black F-150 and began making our way down a patchwork of ranch roads that would bring us to Jurassic Mile. The faded greens of prickly pear cactus, juniper, and sagebrush covered the red hills in the distance, where scorpions and rattlesnakes scuttled in dusty shadows. These venomous creatures, I was warned, would pose the most immediate threat to us at the dig. But during the time when the strata in those hills were first accumulating, the most immediate threats would have been far more conspicuous and would have sent everything else creeping into the shadows. While the herbivorous sauropods we would be digging up would not have been particularly menacing, they would have coexisted alongside more bloodthirsty carnivores.

While Joe navigated from the driver's seat, I asked Susie how she got into studying dinosaurs in the first place.

"This is a true story, even though it doesn't sound like one," she told me, not skipping a beat. At six years old, her grandfather asked her what she wanted to be when she grew up. She had spent family holidays collecting ammonite fossils on the southern coast of England, and loved passing them between her fingers and imagining the worlds they had inhabited. But she was also very taken by royalty.

HEAT

217

"I was wavering viciously between scientist and princess," she said.

Her grandfather, who himself was trained as an electrical engineer, gently nudged her toward science. He knew she loved dinosaurs, so he recommended she become a dinosaur scientist. "I was like, 'okay, sounds good,'" Susie said, looking out the window. "And that's what I did."

In 2015, at the age of thirty-four, she led a team that described the most complete stegosaur skeleton ever discovered and, in so doing, earned the informal title of world's leading expert on stegosaurs. When I asked how she felt about this moniker, she brushed it off, saying it's not hard to be the best at something when only six people do it.

As we approached a barbed-wire fence, Joe put the truck in park and Susie popped out to open the gate. While she fiddled with the latch, Joe turned around to face me with a wry grin. "Susie locked herself on the wrong side the other day," he told me.

Moments later, when she returned to the truck, Susie recounted the same story, with the humility of someone who doesn't take themselves too seriously despite their many accolades.

"I kept jumping out and shutting the gate with great, enormous amounts of physical effort," she told me, "only to find I was on the wrong side."

A woven bracelet dangled from her wrist that matched one that her six-year-old daughter, Amber, wore back home in London. Amber's favorite bedtime stories at that time came from two dinosaur encyclopedias that had been gifted to the family. Though Susie often prefers to read pretty much anything else at the end of a work day, she usually relents, but sets a limit of five pages per night.

We carried on chatting for a bit when all of a sudden, for the first time in close to an hour, a splash of color appeared on the horizon. The royal blues of the Mission Jurassic portable restrooms. If someone were really desperate to find Jurassic Mile, they could pretty easily locate it by way of these toilets.

We pulled past them, parked among several other trucks near the top of a small hill, and stepped out into the Mesozoic.

STRATA

+ + + +

THIS BONE-DRY EXPANSE OF badlands once sat slightly closer to the equator and dripped with meandering streams and muddy floodplains that emptied out into the now-extinct Sundance Sea. Mountains and volcanoes rose up in the south and west, and an ocean sat some 2,000 kilometers to the north. A Mediterranean-like climate offered up dry summers and wet autumns and winters that sustained the vegetation the dinosaurs subsisted on.

The sauropods whose bones we would be digging up lived roughly 150 million years ago, when flowers had yet to spread across continents, and the palettes of landscapes mostly consisted of browns and greens. At certain times, rainstorms had sent channels surging over the banks of the rivers where those dinosaurs congregated. Tendrils of silty water crept onto dry land and left behind drapes of fine sediments that those dinosaurs would muck through as they grazed.

Sometimes, dinosaurs would become sick, or be gnawed on by a carnivore, or get caught in a current, and those dinosaurs would die. Their bodies would collapse into the silt where more tendrils of water would bring more sediment to shroud them in layers that slowly accumulated to the thickness of a bedsheet. Over time the bedsheet grew as thick as a brick, and then a cinder block, and then a skyscraper until eventually the layers reached hundreds of feet high. Flesh decayed and bones remineralized, but the hard structures remained. As time wore on, the land above those remains eroded away, bringing them back up near the surface of the crust where the Mission Jurassic team was working to excavate them.

Once we parked and gathered our belongings, the leaders of the team from the Children's Museum of Indianapolis offered me a tour and a safety briefing. They warned me of snakes, scorpions, and scorching sun, along with the threat of thunderstorms that could appear suddenly and make the dirt road out impassably muddy. The team had already cut one of their days short earlier in the week due

HEAT219

to rain and hail, and the forecast for the next few days threatened more of the same. At the moment, though, a light breeze and clear blue sky promised a comfortable morning.

The excavation spanned two quarries that were separated by a few shipping containers filled with buckets of shovels, brushes, and other neatly organized tools. A bright yellow flag flapped from the top of one container, the wind ruffling an image of a sauropod silhouetted against palm trees.

Once I was properly oriented, I joined the dozen or so people who were already busy digging. Some were shoveling large loads of debris into buckets that they dumped off the side of a cliff while others huddled around with spatulas and precision knives, scraping away at bone.

I stood observing the scene at first, not wanting to intrude. But before long, a woman named Indy, wearing a Pokémon hat, handed me a shovel and asked if I wanted to join in. I held my reporting notebook and audio recorder close to my chest, unsure how to respond. On the one hand, I knew my way around rocks. On the other hand, I had never excavated anything delicate or singular like a dinosaur bone. All the geological samples I had ever collected were comfortingly plentiful. If I broke or lost a piece of limestone or sandstone, I could drop my hammer down and collect more. The same was not true of this fossilized sauropod, a 130,000-pound being who had lived and died and then, against many odds, returned to the surface of the planet more than 100 million years later. I wasn't about to be responsible for somehow messing it up.

But as I watched everyone else become ever sweatier as they dug away beneath the brightening midmorning sun, a twinge of guilt— and a nudge from my inner child—pushed my inhibitions aside. I grabbed a trowel and joined in.

The loose gray dirt contained pale chunks of long, angular rocks that, to my eye, looked just the way I would have expected fragments of bone to look. I wasn't sure how to tell what from what or what I was doing, but I squeezed into a circle of students where Joe was working and asked for their guidance. We kneeled together on foam pads and scraped away at a pile of debris that had been reliably producing bone all week.

To tell the difference between a bone and a rock, some of the more experienced students told me, you look for a reddish hue and for striations similar to those you'd find in petrified wood. If you still aren't sure what you've found, you drag your tongue against it and suck a little bit. If it feels sticky against your tongue, you have found the porousness of bone. If it feels smooth, it's just a rock. This, I was told, is called the lick test.

Before long, Joe's spatula scraped against a reddish, striated chunk the size of his fist. The other students cheered a bit, but soon returned to their side conversations and continued pulling away at the ground with their tools. I couldn't see Joe's expression from behind his sunglasses, and nothing in his body language suggested he was particularly excited. But I knew that, like me, this was his first dinosaur dig, and I couldn't help but wonder how it felt for him to find his first bone.

After he brushed away for a few more moments, he admitted that he was trying to play it cool but was delighted. "This is like my childhood dream," he told me in a low tone, still hunched over the bone and brushing away. A grin widened across his face. "This has always been my goal. Pretty much this second has been my life goal."

Over the next hour or so, I learned that most everyone else in that circle of students had some variation of this same backstory. They had each grown up fascinated by dinosaurs and had never let go of their dream of digging them up.

As I got to chatting with Susie, later on while digging together, I asked about her own interest in dinosaurs, and whether it had waxed or waned over the years she had been studying them. She still finds herself drawn to new discoveries, she told me as she brushed away some dirt, but her deeper interests have shifted increasingly to the strata around the bones. "It's more about understanding the environments," she told me. "The bigger questions, rather than

HEAT 221

just the animals themselves." Those bigger questions are the ones that make dinosaurs relevant to our lives today, as we grapple with global change.

She was doing the heavy lifting required of all the team members of Mission Jurassic, and was curious what new fossils they might turn up. But she was ready to get out into the surrounding ledges of sandstone and siltstone to begin mapping out their broader setting. As this part of the Morrison had been studied far less than other regions, it would provide valuable fresh material to add to her chronology—and to our understanding of the distribution of animals in that hothouse world.

"It is far more than a lifetime's work to study the Morrison," she told me as we continued scraping away at the ground. An increasing body of evidence suggests that the northern Morrison fauna appears to be slightly different from those found in the south. She and her colleagues don't know if this represents a difference in time or geography or both, but her chronology will help them figure this out. If they find that it represents a change over time, then they can begin to ask questions about evolution and what environmental conditions may have forced that shift.

I interrupted our conversation to show Susie a reddish chunk of something as thick as my wrist that had just appeared beneath my trowel. My throat tightened as I swept over it and found more.

"That's a beauty," Susie said, leaning over to take a closer look.

I considered performing the lick test, but I didn't need to. I had found my first dinosaur bone. I touched my hand down and felt the echoes of those sauropod feet hitting the warm Mesozoic earth.

✦ ✦ ✦ ✦

AFTER A COUPLE DAYS at the dig, we ducked away from the quarries to finally spend some time in the surrounding strata.

We descended the hill that we had been throwing debris over and found our way to the base of the Morrison Formation a couple hundred meters below. As the Sundance Sea retreated over time, a broad muddy floodplain with meandering streams emerged in its absence. This was the habitat the sauropods we were digging up had tromped through, eating a daily load of something like one ton of ferns, mosses, horsetails, and other vegetation that grew along the river banks. They swallowed as they grazed, not pausing to chew with their flat teeth but gulping down a rock every once in a while to aid in digestion just as birds—the planet's last remaining dinosaurs—do today.

Near the base of the Morrison, we found a loose fragment of something reddish and striated sitting among the pale rubble at our feet. "Dinosaur," Joe said, placing the fragment in my hand.

"How can you tell it's from a dinosaur?" I asked.

"Unless it's weeny," Susie chimed in, "you can be safe." Everything that lived on land and that grew larger than a meter in size during the Late Jurassic, she reminded me, was a dinosaur.

Since it had been lying loose in the rubble, detached from the rest of the skeleton it had come from, the bone fragment was worthless from a research perspective. It was, as they called it, a "leave-it-asaur," or, in this case, a "souvenir-a-saur." They offered to let me hang on to it, so I stowed it in my backpack as a talisman to help carry us through the rest of the day.

The Morrison stretched several hundred feet above us, in the greenish-gray hillside we had just carefully walked down. Having planted our feet firmly at its base, we began our clamber back up the hill. Susie and Joe draped their hand lenses around their necks on long lanyards, grabbed their rock hammers from their packs, and set to work collecting their stratigraphic log.

The activity is slow but rhythmic. Extend the tape measure; note where you are in the rock face and how far you've come from the previous layer; knock off a piece of the layer with your rock hammer; bring your hand lens close to your eye and bring the piece of rock right beneath the lens; note the size and shape of the sediment, and the quality of its layers. Jot down notes, confer with your

HEAT

field partner to confirm your interpretation of your observations, and then move on to the next layer directly above. If a juniper bush or other obstruction appears in the way, skirt to the right or left in a straight line to find the next well-exposed area and proceed upward, forward in geological time. Keep an eye out for scorpions.

Susie scribbled notes on a clipboard while she and Joe shouted descriptions and measurements back to each other.

"Coarse," Susie called from where she stood.

"Shelly bits," Joe responded from a little higher up.

In between their announcements they paused to drop their rock hammers and hack off chunks of the crumbly stone to look at fresh surfaces beneath their hand lenses. Then they shouted more words to each other.

"Dirty, filthy," Susie called.

"Glauconite," Joe responded.

Every descriptor was code for some characteristic of the environment that the rocks formed within. I was familiar with some words. I knew that *coarse* (as in large-grained) often reflected a high-energy environment capable of moving large grains, like a fast-flowing river. I was less familiar with other words, such as *glauconite*. That, Susie told me, referred to a type of green, fine-grained clay mineral that forms in saltwater and is indicative of a marine or brackish environment. *Dirty, filthy,* another term I was not familiar with in this context, referred to the presence of dark mineral fragments that "dirtied up" the otherwise beige and white quartzes and feldspars that typically dominate sandstones. These strata often represent environments close to a source of erosion—a shore face, say, rather than the middle of the deep sea. For now, though, Joe and Susie weren't drawing any such conclusions. They were simply recording what they found, line by line, pulling together the barcode of this part of the Morrison to make sense of it later on.

"Crossbeds," Susie called, and sketched out a symbol with diagonal lines on her clipboard. She had found a knee-high band of sandy layers that were all slanted at an angle diagonal to the otherwise flat strata above and below. They were the cross section of a large ripple of sediment that likely formed on some sort of small dune at the edge of a body of water, Susie said. She measured the angles of the tilting layers and called out her readings to Joe. As with every line in their barcode, these individual measurements didn't reveal much on the spot. But when linked up with other measurements of crossbeds from around the region, they might help indicate the direction of water flow.

"Planar lamination," Susie called, clambering up the dusty section.

This was code for flatter layers that can either form in very fast flowing conditions or under conditions too slow for any crossbeds or ripples to form at all.

A bird cooed and brought us back to the present.

Another knee-high band of ripply layers sent us back to the Late Jurassic.

"It's a good lat workout," Joe called as he dropped his hammer again.

They gathered together to look at something that they thought might be an oyster fossil, then concluded that it was not an oyster fossil.

"We will not be offended if you decide you want to go and do something far more interesting," Susie told me. Not a chance. I had learned to appreciate each word in each sentence of the strata, to relish in the slow absorption of ancient environments.

"It's not a spectator sport," Joe added.

Spectator sport or not, a film crew of two women from the communications office of the Natural History Museum, London trailed behind us to record an Instagram Live clip of what Susie and Joe were up to that morning. The hashtag #missionjurassic had drummed up some followers across social media platforms that week, and they were trying to harness that attention to introduce their research to a broader audience.

The women fiddled with the angle of their camera to ensure that they weren't revealing any identifying features on the horizon. Then Susie smoothed her hair beneath her hat and faced the camera.

HEAT

225

"So," one of the communications officers said, "there are no dinosaurs right here right now. Why are we not digging right now?"

"Well, today I am out here at the bottom of the Morrison Formation," Susie said, "and I am trying to understand much more about the environments that the dinosaurs were living in. And in order to do that, we need to study the rocks in a lot of detail. So that's what we are doing today."

"Can you show us a little bit of what you have been up to?"

"Yeah, so basically what we do is we just have a look at each layer," she said, pointing out the green-gray rocks around us. "We have been bashing off a bit of each layer and we just look at it in a lot of detail. We make observations about the grain size of the rock and any structures that we can see, whether we can see any fossils like shells, or bits of bone, or bits of plant material. And by looking at this and putting all of this together, we can understand how the rock formed, what sort of processes caused the rock to form. And that allows us to understand the sorts of environments that the dinosaurs were living in." She went on a bit more about her aspirations for the week before signing off.

"Well done; really, really good," the camerawoman said when they finished up.

Susie, back at it with the stratigraphic log, hacked off a bit of a sample and put it in her mouth. "Silty," she said through her teeth. A different type of lick test, one to feel for the grain size of sediment.

I grabbed a chunk of loose rock a bit higher up and put it not to my tongue but to my nose. The smell was so familiar, the dank scent of a musty lake. "I think you're spot-on," Susie told me.

It's this visceral tangibility of deep time that grounds me most when I feel pummeled by the instability of the present. So much and so little has changed since that Mesozoic lake lapped in the wind of that hothouse world. We can't touch time, but we can see, smell, feel and—if we're up for it—taste its relics.

About two-thirds of the way up the hill, the walkie-talkie

chirped on. "You guys need to keep an eye out," we heard a voice say, "because there is this front coming from the southwest, and there is a good chance that it will hit us."

We rogered that and scrambled up the hill to make our way back to the shipping containers for lunch. But our minds were still on the stratigraphic log.

"It's very relaxing," Susie said as we crunched along the pebbly ground, unfazed by the darkening clouds. "I love doing it."

"You lose time," Joe said. "It's mindfulness."

You breathe in and you breathe out. You remember how unlikely your life is, and yet here we are. In this moment, now.

✦ ✦ ✦ ✦

THE WEATHER HELD OFF through lunch, so Susie and Joe descended the hill again to continue with their section while I stuck around in the upper quarry to chat with Paul Kenrick, a paleobotanist from the Natural History Museum, London.

"You have to scrape off the Anthropocene to get down to the Jurassic," Paul said as he dragged a tool over the earth to get to the bone underneath. By the Anthropocene, he meant our current geologic period, the one defined by humans.

He was on-site mostly for the possibility of finding plant fossils that might help reveal more about the Late Jurassic terrestrial ecosystems.

"I hate to say this in front of dinosaur people," he told me, "but plants are far more important than dinosaurs."

Victoria Egerton, one of the dig leaders from the Children's Museum of Indianapolis, laughed and nodded her head.

"I completely agree with you," she admitted.

"When you think about it," Paul went on, "the plants are shaping the environment that the dinosaurs are living in. In fact they are shaping the dinosaurs themselves, the way dinosaurs look."

Long-necked, large-bodied herbivores like sauropods were built that way to access and consume large volumes of vegetation. Their long necks allowed them to graze tall canopies, their dentition allowed them to pull down branches, their gigantic torsos provided the room needed to metabolize all those greens.

Through the Mesozoic, the sway of plants on the shape of animals only became more complex. When flowers bloomed for the first time toward the end of the Mesozoic, pollinator eyes coevolved to detect their bright colors. The emergence of fruits that changed color when they ripened added another layer of complexity that generated more optical sophistication. "They are saying to other animals, 'it's now time to disperse my fruits, come and get it,'" Paul said. "They do that by changing color."

This, he told me, likely played a role in the evolution of trichromatic vision in primates—the ability to see reds, greens, and blues that combine together to generate all the colors on the spectrum of visible light. Most other mammals are dichromatic, only capable of perceiving greens and blues. Cows, for example, only see what they need to see to find grass to graze. Marine mammals, who are monochromatic, evolved to see shapes rather than colors to find their food.

"So it's really plants that are controlling this sort of stuff," Paul said, "and animals are just suckered into it."

"I for one welcome our plant overlords," someone else chimed in, earning laughs from the group.

But plants, of course, are not autonomous. They grow at the whim of the climate and continents beneath them. Their overlords are the other components of the Earth system. Learning how the global climate of the Mesozoic affected the distribution of vegetation, and in turn animals, will offer a whole slew of other insights about how the Earth system functioned back then that could help tighten up predictions of what's to come as the planet continues to warm up today.

"To me, it's about understanding the nature of the Earth system and how the feedbacks work," Paul said. "That is informative for understanding what our impact might be for the future."

He's not only thinking about wild animals and plants, but human society and how it will respond to the stressors it has sown. As the

Earth system spirals and generates panic and unease, human constructs like political systems erode as well.

Even amid the turmoil, Paul and Victoria argue that the study of the natural world must go on.

"If you just stop looking at it and asking those questions, then your understanding of it kind of solidifies," Paul said of the world we live in.

"As the science changes, as society changes, our view of what we can learn from the past changes," Victoria added. The way we understand the rock record now and the questions we ask of it have evolved quite a bit since Hutton and Lyell and Darwin, and will continue to evolve for a long time to come. "There's so much to be discovered."

Centuries ago, geologic discoveries were viewed more as curiosities with little relevance to the modern world. Today, we understand not only their relevance, but the urgency with which we need to consider these insights to make broad-scale decisions about how we operate as a society.

The clouds were beginning to thicken again. Victoria jogged up to the top of the hill to get a sense of how long we had before the rain hit, and decided we should cut the day short.

Before we left, a hubbub formed around one of the shipping containers.

"It's a femur," Susie said, holding a fragment the length of a thumbnail.

Based on its size and curvature, she thought it may have come from a theropod, a group of carnivores that includes *T. rex* but also species as small as chickens.

"The small stuff is less well known; it's rarer," Susie said as others huddled in to get a closer look. "Everyone likes something with pointy teeth, so that's nice. It shows that there are other things in here."

Rain began spotting the trucks, the roofs of the shipping containers, our pants and bags. We piled into our caravan of vehicles and headed out before the road became too slippery to pass, rattling

HEAT

229

over beds of undiscovered bone. Those bones would bring the team back the next day, but the strata surrounding them would bring the bones to life.

✦ ✦ ✦ ✦

YOU CAN'T RUSH STRATIGRAPHY. You must cultivate a practice of patience, a capacity to know you're probably wrong. A willingness to keep going anyway, to keep breathing in and out and keep noticing.

Susie has made significant headway with her chronology, but she will continue working to fine-tune it further and bring those Mesozoic landscapes into sharper focus. Beyond her physical observations from the field, she is collecting a suite of other measurements to help correlate layers across the different stratigraphic logs and begin to build out the ancient landscape in three dimensions. She's looking at changes in the ratios of different elements in the rocks (chemostratigraphy), as well as changes in the orientation of magnetic mineral grains (magnetostratigraphy). She's collecting zircons, where possible, to measure absolute ages of the strata.

"You need a huge, multidisciplinary toolkit," says David Eberth, a geologist in Canada who compiled a chronology similar to the one that Susie is pulling together, but in younger rocks across a smaller area in Canada's Dinosaur Provincial Park. It was the success of that project that inspired Susie to try to do the same within the Morrison Formation.

With his chronology, Eberth and his colleagues were able to determine that differences in crest shapes in hadrosaurs in different locations represented evolution through time rather than differences between males and females who coexisted at the same time, as they had previously thought. Having clarified this point, they can now explore what conditions may have catalyzed that change in crest shape. Perhaps it was environmental, perhaps more biological, or a mix of the two.

Only when multiple lines of evidence match up in layers across multiple logs can Eberth or Susie begin to connect the dots in time and space and gain their bearings enough to place the dinosaurs within the frameworks they are building—and from there, gain insights about the distribution of large-bodied animals across a warmer version of Earth.

Susie's work in the Morrison is just one piece of the Mesozoic puzzle, capturing a snapshot of life on land in a single region of the planet. But what were the oceans doing at this time? How were other nonliving components of Earth, like wildfires, adjusting to the warmth? This is where the work of Karen Chin, Ian Glasspool, and hundreds of others like them comes in. They all need each other. Each of these different silos of research feeds into a more complete picture of the Earth system at this time. We won't ever know exactly how the planet looked or felt over the millions of years of the Mesozoic. But as we inch closer to a clearer picture, we deepen the intimacy with which we know Earth and its capacity to withstand heat. The more we know, the more we care.

For Eberth, the value of stratigraphic work like his and Susie's lies beyond understanding the animals and the environments they lived in during the Mesozoic: It is through this study of deep time that we gain the capacity to point out differences in the rates of change we're experiencing today as compared with those of the past. "You will encounter climate change deniers who will throw out the comment that climates have changed in the past, and they are absolutely correct as far as that statement goes," Eberth says. But what gets lost in that statement is the nuance of how rates of change dictate how the Earth system responds to change. "Ecosystems can adapt reasonably well at a slower rate," he says. "But the minute you start ramping the rates up, that's when we start seeing extinctions."

If the pace of warming during the end-Permian mass extinction was orders of magnitude slower than modern global temperature warming, and was enough to mark the single worst mass extinction in Earth history, then we should heed that warning.

HEAT

231

And yet. The resolution of the rock record is only so strong. The pace of change in the past may have been faster or slower than current estimates suggest, but we can't know for certain because we only have so many data points to pull from.

The only way to gain more clarity is to continue pulling sentences from the strata. To assign more ages to more layers, increase our resolution, and improve our understanding of what the proxies are telling us.

✦ ✦ ✦ ✦

THE END OF THE Mesozoic crops up as a thin layer of pale clay. Those fine, dusty particles contain loads of the element iridium in concentrations orders of magnitude greater than the surrounding bedrock. Together, the clay and the iridium say that the hothouse Mesozoic moment ended in a crash.

An asteroid half the size of Manhattan barreled down and hit Earth off the coast of what is now Mexico's Yucatan Peninsula, vaporizing the land beneath it and sculpting a crater more than one hundred miles wide. The aftermath of the impact began as violent firestorms and tsunamis, but persisted far longer as an intense period of cooling and darkness. The dissipation of vaporized rock particles into the stratosphere created a shield that blocked incoming light and heat from the sun. Plant growth slowed. Food sources dwindled. Air temperatures dropped. Even the tropics may have hovered around freezing for some months.

Out went the non-avian dinosaurs, the marine reptiles, the ammonites. Three-quarters of all animal species perished during this end-Cretaceous mass extinction, the event that drew the Mesozoic to a close and sent shockwaves through the entire Earth system. Whereas other mass extinctions have stretched over the course of millions of years, this one stretched at most on the order of tens of thousands of years, and possibly far shorter. One estimation suggests that extinctions could have begun within nine months of the impact.

232 STRATA

Ice caps expanded. Cold, dense seawater sank to the seafloor, nutrient-rich waters plumed up to replace them at the ocean surface and fertilized algal blooms that poisoned currents with toxicity.

Over time, though, life did bounce back. Flowers blossomed and bloomed. Grasses sprouted. Animals licked their wounds and crept back out from beneath the shadows. Mammals grew and filled the monstrous void left behind by the dinosaurs. Placentas proliferated, and with them the hairy beings that would eventually evolve into us.

If that asteroid had landed just a few minutes later in water just a few hundred meters deeper, everything could have been different. Less rock may have vaporized, less solar radiation may have been blocked. Sauropods and stegosaurs and tyrannosaurs and all the rest of them may have clawed their way through and come out on the other side, never opening up the space for those small, hairy beings to rise to dominance.

We grew from that devastation, that piece of space debris thrown randomly down to Earth. That wreckage brought with it the possibility of bouquets and dance parties and fresh-baked pastries; orchestras and paintings with pigments in heartbreaking hues. It brought the possibility of narratives, of stories passed down through mouths and later through hands and ink and paper and now silicon. Of people who could weave stories from stone.

This hits me not only when I'm sitting with strata but when I'm out in the world in a wholly human experience. I'm at a concert at a small stadium, my shoulder brushes the arm of the man sitting next to me. He slurps soda from a plastic straw and bobs his head as he mouths the words to nobody. Everyone is sweaty and we all know the song. When the stage lights flicker from fuchsia to yellow to alien green in tempo with the bass, we scream and cheer because we get to be here. The lights flash again and for a moment, in the darkness, I'm not here but floating through the depths of an Ediacaran sea. I'm swimming through the brine of those sightless, earless, boneless beings. I'm remembering the ice ages that they sprang

HEAT

from and the mudlessness of land in those days. I'm remembering the lunglessness and soundlessness and now I'm remembering to breathe and turn back to the music.

But I can't take my mind off it. How, after all those epochs, after everything Earth had been through, the thing that made this moment possible may have just been that random piece of space junk.

We can touch its impact, that layer of clay thinner than your pinky. Rest hands on the calamity that paved the way for us.

Epilogue: Us

AT SOME POINT WHEN I WAS IN HIGH SCHOOL, IN THE EARLY 2000s, my family crammed into our minivan and drove down to my father's childhood home in New Jersey to help move our grandmother into a nursing home. Her dementia was getting worse by then, and we needed to clear out her house and put it on the market for the first time in more than half a century.

The house was a drafty old Victorian filled with mattress-sized heaps of decades-old issues of *National Geographic*, a refrigerator full of expired food, stethoscopes from the years that my grandfather ran his medical practice out of their home, and knickknacks scattered in many dusty corners.

Overwhelmed by the chaos, I escaped into the attic. I was relieved to find mostly empty space, save for a chest that sat in a beam of sunlight streaming in through a dirty window. Inside, under a layer of what looked like my father's childhood homework, I unearthed a stack of envelopes. I slid one out from the rubber band that held the stash together and found several pages scrawled with professions of love, ending with my grandfather's signature. I thumbed through a few other envelopes and realized that they were a collection of love letters he and my grandmother had exchanged while they were courting during World War II, a decade or so before my father's birth. The two would have been in their late teens and early twenties, just a handful of years older than I was at that time. On one lengthy page, my grandfather promised to care for my grandmother until she grew old, until she was crusty with drool falling from her mouth.

236   STRATA

I had only ever really known that version of my grandmother. The papery-skinned, fragile woman with a gravelly Brooklyn accent and head of white hair. I knew even less of my grandfather, who had passed away more than a decade earlier. But with my finger pressed against those inky lines, I could imagine, for the first time, younger versions of those humans. I felt like I could relate to them, like I had something in common with them. Their words saturated the hues of the string of events that landed my father in the world, and that would eventually send me up into that sunbeam in the dusty attic of their home.

That warmth, that sense of connection that I found in those letters felt not unlike the warmth I now receive from lines in stone. Strata are, in a certain way, love letters left behind by an aging Earth. They remind us where we came from. That we live our lives in a recycled world, and have created nothing from scratch. Even as the planet ages and grows sick, its stories persist as constant reminders that return us home. Reminders that we are the product of a system that has been humming for 4.54 billion years, and that we carry its beginnings in our bones.

✦   ✦   ✦   ✦

WHEN RACHEL CARSON PUBLISHED *The Sea Around Us* in 1951—a book that reads to me like a love letter to the earth—the scientific community was already well aware of changes unfolding in the global environment. They had already established beyond question, Carson wrote, that a clear change in the Arctic had set in by 1900 and had become "astonishingly marked" by about 1930.

"The frigid top of the world is very clearly warming up," she wrote.

That book sold some 250,000 copies the year it was published and sat on the *New York Times* bestseller list for eighty-six weeks. It won the National Book Award in 1952, and would ultimately sell more than 1 million copies and be translated into twenty-eight languages.

EPILOGUE: US   237

But here we are, more than seventy years later, in many ways none the wiser. Still struggling—or refusing—to come to grips with the facts she shared, even as they glow with eerie warmth all around us.

As I write this, Earth is spinning through its hottest year on record since humans first started collecting such records in the mid-1800s. The humans most at risk from this warming are those with the fewest resources to cope with it, and who have contributed least to the extractions that brought us here. One 2023 study found that between 1980 and 2010, the poorest 20 percent of the planet's population experienced an increase of between 24 to 48 percent more drought-to-downpour events, or weather whiplash as the study's authors call it. Such extremes have decimated agricultural land and destroyed homes, and continue to threaten lives and livelihoods for hundreds of millions of people each year.

We speak of all this today as if it is a new problem, but it's one that has been stewing for many years now. We're simply no longer able to look the other way.

Carson wrote of Baltimore orioles, Canada warblers, and a whole collection of other more southerly bird species flocking to Greenland for the first time by the 1920s. She described sea ice in the Russian Arctic shrinking by a million square kilometers between the years 1924 and 1944. She wrote of ships traveling farther north than they ever had in the 1930s, and the coal shipping season in Svalbard expanding from three to seven months by the 1940s, propelled by a warming Gulf Stream. Greenland's barely existent cod fishery grew to 15,000 tons of catch per year by midcentury, while glaciers in Europe, North America, South America, Asia, and Africa were all showing signs of retreat. Oats sprouted more vigorously in Iceland as spruce and pine forests marched farther north in Scandinavia, punching through the tundra with new lush growth—all by the 1940s, less than 200 years after the onset of the Industrial Revolution.

"The evidence that the top of the world is growing warmer is to be found on every hand," Carson wrote.

238   STRATA

The cause for that warming was yet unknown, but the trajectory
was clear.

"The long trend is toward a warmer earth; the pendulum is swinging," she ended one of her last chapters.

✦ ✦ ✦ ✦

RACHEL CARSON PASSED AWAY in 1964, not long before Elso
Barghoorn published his second paper on the Gunflint Chert and
Brian Harland published his thoughts on the Great Infra-Cambrian
Ice Age. Right in the midst of the plate tectonic revolution, when
Paul Hoffman was being told not to study tectonics but doing it anyway, and Stu Roscoe was careening through the Canadian Arctic
finding uranium grains in ancient riverbeds. A whole new swell of
planetary understanding was rising just as Rachel Carson was slipping away. She never had the privilege of knowing Earth with the
intimacy we get to today.

Since her passing, we have discovered more about Earth's past
than Carson could have easily imagined. We have learned that oxygen was all but absent in the atmosphere for more than half of Earth's
existence, and that the eventual rise of this gas allowed multicellular
life to evolve. That a series of global glaciations may have helped
explode and diversify that complexity into beings with eyes and lips
and bones and brains. That these complex beings lived entirely in
the oceans for hundreds of millions of years, until the rise of mud
on land finally ushered in the rise of floodplains where the first terrestrial communities sought refuge. And that as those communities
lushed up and warmed during the Mesozoic, hundreds of millions of
years later, they generated deposits that mirror what's to come of our
future as we barrel into a warmer world.

By pushing ourselves to the edge of crisis, we have found ourselves nose to nose with our origins. Our planet seems to be telling
us to take a look back. To remember how unlikely each of our lives
is, and yet here we are.

EPILOGUE: US   239

✦ ✦ ✦ ✦

WE NOW KNOW THAT the sun is a little less than halfway through its life. That within about 5 billion years, it will die and vaporize Earth and all the materials that make up our bodies, throwing every trace of everything that has ever transpired here back into the cosmos where they came from.

But that's still a long, long way away from now. Until then, Earth will carry on shedding bits of itself to make room for new growth. It might, in time, shed us. We will, no doubt, shed many other lifeforms in the meantime, and for this I have no words. But even as the characters change and the conditions shift, Earth will still be here, a solid sphere whipping around the sun, just as it has always done.

The gift of geology is the chance to seek refuge in this constancy, in the gravity of the arc of time. When I walk the rocky shoreline near my home, I don't see random stones thrown about but a montage of stories and events that intertwine directly with our present and our future.

I discovered recently that some Mesozoic basalt crops up on the far end of a coastal park not far from where I live. The magma had spewed up through fractures in bedrock that had split open when Pangea pulled apart, like taffy torn to pieces. Earth's insides took advantage of those brittle gaps and shot up through them to form basalt dikes, linear stretches of rusty red rock that appear all across the coast of southern Maine.

The dike that runs through that particular coastal park cuts through gray quartzites and rusty phyllites that had once composed a deep seafloor during the Silurian. By the time that Pangea was pulling apart and spewing up lavas, the Silurian was already more than 250 million years in the past, as far back as the Mesozoic is today. There's deep time even in deep time, endless onion skins of antiquity.

With the help of a geologic map, I found that dike in a gully. I sat there with a friend for a while, facing out toward the Gulf of Maine that splashed toward us with an incoming tide. We drank herbal tea out of a thermos and watched clouds drag shadows over the warming sea. Gulls floated above us as we spoke of ancient magmas and the ends of time. What might come of us when the sun dies, if we make it that long.

I mindlessly fiddled with a piece of the basalt as we spoke, dragged that bit of the Mesozoic beneath my fingernail. Blew its iron dust away into the wind.

We wandered along the shoreline and found a large boulder where visitors had carved their initials. What at first looked like recent scribblings we realized were actually more than one hundred years old, with dates etched from the early 1900s next to the names. So recent, by certain measures, but also so far in the past. Whoever those humans were, they were born before the first atomic bomb left radioactive fingerprints within tree rings and seashells. Before our microplastics filled the biosphere, before our asphalts sealed the lithosphere. Before the hydrosphere acidified with our combustion of fossil fuels in the atmosphere. Before humans laced their trace of activities throughout all facets of the Earth system, laying claim to the next rising geologic epoch: the Anthropocene, the Age of Humans. Or, as others have come to call it, the Capitalocene—an age defined not by all humans, but by humans participating in the extractions and consumptions of capitalism.

Behind the inscriptions in that boulder, we found a long and narrow gap in the Silurian rock that was just wide enough to fit our bodies within. I imagined that maybe a basalt dike had once shot through there but had since eroded away. I scrunched my legs up to my chin and nestled in, hugged on both sides by the Silurian. My friend did the same, facing me, and we sat there together, held by those rocks, and kept talking about history on all scales.

Water slapped the rocks beyond us. The tide inched farther ashore as the hours ticked on and we kept talking. Waves crested and crashed, falling and regrouping again, with the tenacity of a toddler just learning how to walk. As if those waves hadn't been moving

EPILOGUE: US

with that motion for more than 4 billion years, as if they didn't know what they were in for.

If there's one thing we can say with certainty has remained constant since at least the Archean, it's the persistent tug of water against rock and the erosion that comes with it. The breaking down of Earth's skin and bones to make room for something new. The motion is at once unchanging and the most persistent force of change. It is carving down boulders into cobbles into pebbles into sands, silts, clays. It is turning land into dust and sending its debris back to the sea it came from.

By the time the seafloors of today rise up above the oceans as cliff-sides or mountaintops, our individual lives will be specks of dust, imperceptible to the naked eye. The iron in our blood will have pooled back into the earth, all our remains melting within the mantle where we will meet, again, as one.

Acknowledgments

TWENTY YEARS AGO, MY HIGH SCHOOL ENVIRONMENTAL science teacher suggested I take a geology class once I got to college. I had never taken an Earth science class before, and the thought had never crossed my mind. Had it not been for that suggestion, this book would almost certainly not exist. Thank you to the late Bob Oddo for opening that door and recognizing my interest in the geosciences before I had.

I am deeply grateful to my agent, Sarah Fuentes, whose sharp eye has helped shape this book since its inception and whose thoughtfulness and support have buoyed me over the years. Thank you to my brilliant editors, Drew Weitman and Jessica Yao, for championing this project and for improving it beyond measure. Thanks also to Annabel Brazaitis, Yumiko Gonzalez Rios, copyeditor Mike van Mantgem, and the whole team at W. W. Norton for your help getting this book out into the world.

A deep thanks to the incredibly talented Sarah Gilman, whose gorgeous illustrations have added so much to these pages. I am honored to be able to include your artwork in this book.

The completion of this project relied on the patience and generosity of dozens of scientists who shared their time with me to help me understand their research. I am grateful to everyone I spoke with, particularly those who maintained a discourse over the span of years: Woody Fischer, Usha Lingappa, Sarah Zeichner, Andy Knoll, Amy Radakovich Block, Latisha Brengman, Paul Hoffman, Adam Maloof, Ryan Ewing, Paul Myrow, Linda Ivany, Neil Davies, Ian Glasspool, Karen Chin, and Susie Maidment.

ACKNOWLEDGMENTS

Thank you to Bev Johnson, Mike Retelle, Dyk Eusden, Dave Jones, Julie Brigham-Grette, and Ross Powell for introducing me to the geosciences and for your encouragement as I first discovered this field. Thanks to Rob Irion for your steadfast guidance and mentorship early in my career. A special thanks to Rosie Mestel, who assigned and brilliantly edited three stories that originally appeared in *Knowable Magazine* and later helped shape three of the four sections of this book.

I am grateful to the Hewnoaks Artist Residency for providing the creative space I needed to first conceive of this idea, and to Donna Palley and Steven Scudder for providing the quiet space I needed to complete my first draft. Thank you to the Portland Dharma House for years of wisdom, friendship, and support.

I have been lucky enough to find myself in community with many other writers who have helped shape this book and have boosted morale through the ups and downs. Thank you to Chelsea Steinauer-Scudder, Jenny O'Connell, Kathryn Amato, Emily Powers, and Julia Rosen for your warmth and smart feedback. Thank you to Kelly Servick, Kathryn Miles, Brandon Keim, and everyone on the Slackline for your years of camaraderie.

I am grateful to this planet for bringing us into existence. I try to live my days honoring this gift.

Thank you to my family for your encouragement and for sharing your deep curiosity in the world. Your love and support mean everything.

Notes

Prologue

1 **lake that once ran as a river:** Raymond A. Soltero, John C. Wright, and Abraham A. Horpestad, "Effects of Impoundment on the Water Quality of the Bighorn River," *Water Research* 7, no. 3 (March 1973): 343–54.

1 **melting glaciers sent torrents:** Yellowstone Volcano Observatory, "The Grand Story of the Grand Canyon of the Yellowstone," US Geological Survey website, January 16, 2023.

1 **the construction of a dam:** "Yellowtail Dam," Bighorn Canyon National Recreation Area website.

2 **caught somewhere in the Paleozoic:** "Geologic Story," Bighorn Canyon National Recreation Area website.

One

14 **"touches upon tens of hundreds of stories":** John McPhee, *Basin and Range* (New York: Farrar, Straus and Giroux, 1981), 35.

14 **one lone parasite of Chinook salmon:** Dayana Yahalomi et al., "A Cnidarian Parasite of Salmon (Myxozoa: *Henneguya*) Lacks a Mitochondrial Genome," *Proceedings of the National Academy of Sciences* 117, no. 10 (February 24, 2020): 5358–63.

15 **"And so . . . swirling clouds and gloom":** Rachel Carson, *The Sea Around Us* (New York: Oxford University Press, 1951), 22.

16 **weaseled into microbial cells:** James A. Imlay, "The Molecular Mechanisms and Physiological Consequences of Oxidative Stress: Lessons from a Model Bacterium," *Nature Reviews Microbiology* 11, no. 7 (July 2013): 443–54.

16 **made arsenic more soluble:** Song-Can Chen et al., "The Great Oxidation Event Expanded the Genetic Repertoire of Arsenic Metabolism and Cycling," *Proceedings of the National Academy of Sciences* 117, no. 19 (May 12, 2020): 10414–21.

17 **As hard as paleontologists of that time looked:** William J. Schopf, "Solution to Darwin's Dilemma: Discovery of the Missing Precambrian Record of Life," *Proceedings of the National Academy of Sciences* 97, no. 13 (June 20, 2000): 6947–53.

17 **"The case at present must remain inexplicable":** Charles Darwin, *On the Origin of Species by Means of Natural Selection, or the Preservation of Favoured Races in the Struggle for Life* (London: John Murray, 1859), 308.

17 **In the summer of 1953:** The following account comes from William J. Schopf, *Cradle of Life: The Discovery of Earth's Earliest Fossils* (Princeton: Princeton University

NOTES

Press, 2021), 35–70; and Robert H. Dott Jr., "Serendipity and Stan Tyler's Precambrian Gunflint Fossils," *The Outcrop* (2000): 25–26.

18 **the two published a short paper:** Stanley A. Tyler and Elso S. Barghoorn, "Occurrence of Structurally Preserved Plants in Pre-Cambrian Rocks of the Canadian Shield," *Science* 119, no. 3096 (April 30, 1954): 606–8.

19 **published it in the journal *Science*:** Elso S. Barghoorn and Stanley A. Tyler, "Microorganisms from the Gunflint Chert: These Structurally Preserved Precambrian Fossils from Ontario Are the Most Ancient Organisms Known," *Science* 147, no. 3658 (February 5, 1965): 563–75.

19 **a couple months before Cloud:** Preston E. Cloud Jr., "Significance of the Gunflint (Precambrian) Microflora," *Science* 148, no. 3666 (April 2, 1965): 27–35.

19 **"For all of time":** Schopf, *Cradle of Life*, 59.

19 **no oxygen gas in it at all:** Cloud, "Significance of the Gunflint," 27–35.

20 **planet's earliest atmosphere likely lacked oxygen:** Alfred G. Fischer, "Fossils, Early Life, and Atmospheric History," *Proceedings of the National Academy of Sciences* 53, no. 6 (1965): 1205–13.

20 **made its own atmosphere:** Preston E. Cloud Jr., "Atmospheric and Hydrospheric Evolution on the Primitive Earth: Both Secular Accretion and Biological and Geochemical Processes Have Affected Earth's Volatile Envelope," *Science* 160, no. 3829 (May 17, 1968): 729–36.

21 **not algal at all:** Ferran Garcia-Pichel et al., "What's in a Name? The Case of Cyanobacteria," *Journal of Phycology* 56, no. 1 (February 2020): 1–5.

21 **consensus today points to cyanobacteria:** Woodward W. Fischer, James Hemp, and Jena E. Johnson, "Evolution of Oxygenic Photosynthesis," *Annual Review of Earth and Planetary Sciences* 44, no. 1 (June 29, 2016): 647–83.

21 **recent genomic research:** Rochelle M. Soo et al., "On the Origins of Oxygenic Photosynthesis and Aerobic Respiration in Cyanobacteria," *Science* 355, no. 6332 (March 31, 2017): 1436–40.

21 **simpler photosynthesizers made their food:** Fischer et al., "Evolution of Oxygenic Photosynthesis," 647–83.

Two

23 **coauthored a paper with him:** Ryan C. Ewing et al., "New Constraints on Equatorial Temperatures During a Late Neoproterozoic Snowball Earth Glaciation," *Earth and Planetary Science Letters* 406 (November 2014): 110–22.

24 **coined this pivotal moment:** Heinrich D. Holland, "Volcanic Gases, Black Smokers, and the Great Oxidation Event," *Geochimica et Cosmochimica Acta* 66, no. 21 (November 1, 2002): 3811–26.

24 **the term Great Oxygenation Event:** Woodward Fischer, email message to the author, May 2, 2024.

25 **greatest environmental engineers:** Woodward W. Fischer, James Hemp, and Joan Selverstone Valentine, "How Did Life Survive Earth's Great Oxygenation?" *Current Opinion in Chemical Biology* 31 (April 2016): 166–78.

26 **sulfur chemistry of seafloor sediments:** Timothy W. Lyons, Christopher

NOTES

T. Reinhard, and Noah J. Planavsky, "The Rise of Oxygen in Earth's Early Ocean and Atmosphere," *Nature* 506, no. 7488 (February 2014): 307–15.

26 **less than 1 percent of modern levels:** Noah Planavsky et al., "Low Mid-Proterozoic Atmospheric Oxygen Levels and the Delayed Rise of Animals," *Science* 346, no. 6209 (October 31, 2014): 635–38.

26 **suffused with carbon dioxide, methane, and water vapor:** "The Great Oxidation Event: How Cyanobacteria Changed Life," American Society for Microbiology website, February 18, 2022.

26 **oxygenate the ocean:** Richard G. Stockey et al., "Sustained Increases in Atmospheric Oxygen and Marine Productivity in the Neoproterozoic and Palaeozoic Eras," *Nature Geoscience* 17, no. 7 (July 2024): 667–74.

27 **fossils in rocks as old as 3.5 billion years:** William J. Schopf and Bonnie M. Packer, "Early Archean (3.3-Billion to 3.5-Billion-Year-Old) Microfossils from Warrawoona Group, Australia," *Science* 237, no. 4810 (July 3, 1987): 70–73.

27 **thirty locations around the world:** Andy Knoll, interview by the author, April 6, 2022.

27 **so-called "whiffs of oxygen":** Ariel D. Anbar et al., "A Whiff of Oxygen Before the Great Oxidation Event?" *Science* 317, no. 5846 (September 28, 2007): 1903–6.

27 **refute some evidence for:** Sarah P. Slotznick et al., "Reexamination of 2.5-Ga 'Whiff' of Oxygen Interval Points to Anoxic Ocean Before GOE," *Science Advances* 8, no. 1 (January 5, 2022): eabj7190.

27 **"We find these arguments logically flawed":** Ariel D. Anbar et al., "Technical Comment on 'Reexamination of 2.5-Ga "Whiff" of Oxygen Interval Points to Anoxic Ocean Before GOE,'" *Science Advances* 9, no. 14 (April 7, 2023): eabq3736.

29 **later dated at around 1.9 billion years old:** Philip Fralick, Don W. Davis, and Stephen A. Kissin, "The Age of the Gunflint Formation, Ontario, Canada: Single Zircon UPb Age Determinations from Reworked Volcanic Ash," *Canadian Journal of Earth Sciences* 39, no. 7 (July 1, 2002): 1085–91.

31 **"Evidence of our shared origin":** Sherri Mitchell, *Sacred Instructions: Indigenous Wisdom for Living Spirit-Based Change* (Berkeley, CA: North Atlantic Books, 2018), 4.

Three

33 **wealth and friends in high places:** Marcia Bjornerud, *Reading the Rocks: the Autobiography of the Earth* (New York: Basic Books, 2005), 27.

33 **6,000 years since the planet's creation:** James Barr, "Why the World Was Created in 4004 B.C.: Archbishop Ussher and Biblical Chronology," *Bulletin of the John Rylands Library* 67, no. 2 (March 1985): 575–608.

33 **"The result, therefore, of this physical inquiry":** James Hutton, "Theory of the Earth; or an Investigation of the Laws Observable in the Composition, Dissolution, and Restoration of Land upon the Globe," *Transactions of the Royal Society of Edinburgh* 1 (1788): 96.

34 **"a regular succession of Earths":** This quote originally appears in an anonymous article published in *The Monthly Review* 79 (1788): 36–37, as Jack Repcheck notes

in *The Man Who Found Time: James Hutton and the Discovery of the Earth's Antiquity* (Reading, MA: Perseus, 2003), 159.

34 **unknowably long stretches of time:** Repcheck, *Man Who Found Time,* 152–53.

34 **most important book:** Marcia Bjornerud, *Timefulness: How Thinking Like a Geologist Can Help Save the World* (Princeton: Princeton University Press, 2018), 26.

34 **"The ancient people perceived":** Leslie Marmon Silko, *Yellow Woman and a Beauty of the Spirit* (New York: Simon and Schuster, 1997), 31.

35 **span some 4.32 billion years:** "Concept of Time," Hindu Online website, 2010.

35 **4.54-billion-year-old age of Earth:** Brent G. Dalrymple, "The Age of the Earth in the Twentieth Century: A Problem (Mostly) Solved," *Geological Society, London, Special Publications* 190, no. 1 (January 2001): 205–21.

35 **as long as 75,000 years:** Morten Rasmussen et al., "An Aboriginal Australian Genome Reveals Separate Human Dispersals into Asia," *Science* 334, no. 6052 (October 7, 2011): 94–98.

35 **creation of the world as Dreamtime:** "Aboriginal Dreamtime," Artlandish Aboriginal Art Gallery website, 2024.

35 **by certain measures, to this day:** Meghana Ranganathan et al., "Trends in the Representation of Women Among US Geoscience Faculty from 1999 to 2020: The Long Road Toward Gender Parity," *AGU Advances* 2, no. 3 (September 2021): e2021AV000436.

35 **most detailed and expansive geologic map:** "William Smith's Geological Map of England," NASA Earth Observatory website, May 10, 2008.

35 **advent of radiometric dating:** Tess Joosse, "February 1907: Bertram Boltwood Estimates Earth Is at Least 2.2 Billion Years Old," American Physical Society website, January 11, 2024.

35 **development of mass spectroscopy:** John De Laeter and Mark D. Kurz, "Alfred Nier and the Sector Field Mass Spectrometer," *Journal of Mass Spectrometry* 41, no. 7 (July 2006): 847–54.

36 **to advise the Canadian government:** Paul Hoffman, interview by the author, December 17, 2021.

36 **en route to fill up on:** D. C. Findlay, "Incident at Booth River About 1985," *Friends of GSC History* (2007), Series A—Historical Contributions, GSCHIS-A008.

37 **"It leads . . . to a consideration":** S. M. Roscoe, *Huronian Rocks and Uraniferous Conglomerates in the Canadian Shield* (Ottawa: Printing and Publishing Supply and Services Canada, 1976), 2.

38 **pyrite rapidly disintegrates:** Jena E. Johnson et al., "O_2 Constraints from Paleoproterozoic Detrital Pyrite and Uraninite," *Geological Society of America Bulletin* 126, no. 5–6 (May 2014): 813–30.

38 **rest in museum display cases:** Woody Fischer, interview by the author, October 19, 2021.

40 **from about 3.6 billion to 2.5 billion years:** Bor-Ming Jahn and Kent C. Condie, "Evolution of the Kaapvaal Craton as Viewed from Geochemical and Sm-Nd Isotopic Analyses of Intracratonic Pelites," *Geochimica et Cosmochimica Acta* 59, no. 11 (June 1995): 2239–58.

41 **less than 3 percent of Earth's modern surface:** Carol Frost et al., "Creating Continents: Archean Cratons Tell the Story," *GSA Today* 33, no. 1 (January 2023): 4–10.

NOTES 249

41 **77 percent of the world's:** "The Kalahari Manganese Field," Ntsimbintle Holdings website, 2024.

41 **a rare manganese-poisoning disease:** J. A. Neser et al., "The Possible Role of Manganese Poisoning in Enzootic Geophagia and Hepatitis of Calves and Lambs: To the Editor," *Journal of the South African Veterinary Association* 68, no. 1 (July 13, 1997): 4–6.

41 **radiation doses 1,000 times greater:** Feng Liu, Nuomin Li, and Yongqian Zhan, "The Radioresistant and Survival Mechanisms of *Deinococcus radiodurans*," *Radiation Medicine and Protection* 4, no. 2 (June 2023): 70–79.

42 **protect against cancer-causing free radicals:** "Manganese," Mount Sinai website, 2024.

42 **possibly 1,000-fold greater:** Usha F. Lingappa et al., "How Manganese Empowered Life with Dioxygen (and Vice Versa)," *Free Radical Biology and Medicine* 140 (August 2019): 113–25.

43 **"When I saw them":** Usha Lingappa, "Usha Lingappa PhD Thesis Defense Seminar," March 3, 2021.

44 **"These are the communities":** Lingappa, "PhD Thesis Defense."

45 **source of this heat:** Mikio Fukuhara, "Possible Generation of Heat from Nuclear Fusion in Earth's Inner Core," *Scientific Reports* 6, no. 1 (November 23, 2016): 37740.

46 **komatiite that all but disappears:** Tim E. Johnson et al., "Delamination and Recycling of Archaean Crust Caused by Gravitational Instabilities," *Nature Geoscience* 7, no. 1 (January 2014): 47–52.

46 **flooding marine basins with far more water:** Junjie Dong et al., "Constraining the Volume of Earth's Early Oceans with a Temperature-Dependent Mantle Water Storage Capacity Model," *AGU Advances* 2, no. 1 (March 2021): e2020AV000323.

46 **preponderance of underwater volcanoes:** Lee R. Kump and Mark E. Barley, "Increased Subaerial Volcanism and the Rise of Atmospheric Oxygen 2.5 Billion Years Ago," *Nature* 448, no. 7157 (August 2007): 1033–36.

47 **more than twenty different rationales:** Aleisha Johnson, interview by the author, March 6, 2024.

47 **react more readily with oxygen:** Heinrich D. Holland, "Volcanic Gases, Black Smokers, and the Great Oxidation Event," *Geochimica et Cosmochimica Acta* 66, no. 21 (November 1, 2002): 3811–26.

47 **made this case in a 2007 paper:** Kump and Barley, "Increased Subaerial Volcanism," 1033–36.

Four

49 **some trace of Soudan iron:** Ann Wessel, "Minnesota's Soudan Mine Now Produces Awe Instead of Ore," *St. Cloud Times*, September 20, 2014.

49 **donated the mine and the surrounding 1,200 acres:** "Lake Vermilion-Soudan Underground Mine State Park," Minnesota Department of Natural Resources website.

49 **welcomes tens of thousands of visitors:** Dan Kraker, "After Four-Year Hiatus, Underground Mine Tours Resume at Soudan State Park," *MPR News*, May 24, 2024.

49 **Iron deposits like these generally exist only:** Jiangning Yin, Han Li, and Keyan

Xiao, "Origin of Banded Iron Formations: Links with Paleoclimate, Paleoenvironment, and Major Geological Processes," *Minerals* 13, no. 4 (April 13, 2023): 547.

50 **a single liter of ocean water:** Emily Underwood, "The Iron Ocean," *Knowable Magazine,* December 19, 2019.

51 **geochemistry of iron deposits:** Athena Eyster et al., "A New Depositional Framework for Massive Iron Formations After the Great Oxidation Event," *Geochemistry, Geophysics, Geosystems* 22, no. 8 (August 2021): e2020GC009113.

52 **body of water Onamanii-zaaga'igan:** Jaylen Strong, email message to the author, July 10, 2024.

52 **see if those rumors held true:** David A. Walker, "Lake Vermilion Gold Rush," *Minnesota History* 44, no. 2 (Summer 1974): 42–54. The following account of the Lake Vermilion Gold Rush comes from this source.

52 **"The eye takes in a score or more graceful":** Oro Fino, "Golden Letters, Number Nine," *The Weekly Pioneer and Democrat* (October 27, 1865), 5. Available online at: https://newspapers.mnhs.org/jsp.

53 **1,000 acres of the tens of thousands:** "Zagaakwaandagowininiwag/Bois Forte Band of Chippewa," Minnesota Indian Affairs Council website.

54 **received recognition . . . from groups:** "A Vision to Preserve Our History," Bois Forte Heritage and Cultural Museum website.

54 **"Any attempt to disentangle":** Lauret Savoy, *Trace: Memory, History, Race, and the American Landscape* (Berkeley, CA: Counterpoint Press, 2015), 66.

55 **only 105 were Native:** Kelly Kang, "Research doctorate recipients, by ethnicity, race, and citizenship status: 2011–21," *National Center for Science and Engineering Statistics, Survey of Earned Doctorates,* Table 1-8 (October 18, 2022).

55 **Indigenous Geoscience Community:** "EAGER-Indigenous Geoscience Community," National Science Foundation Award Abstract # 2039338.

55 **collaborating with a muralist:** " 'Everything is connected': Mural Inspired by UMD Professor Celebrates Traditional Haida Knowledge," University of Minnesota Duluth website, April 18, 2022.

55 **hovered around 2.7 billion years old:** George J. Hudak et al., "Bedrock Geology of Lake Vermilion/Soudan Underground Mine State Park," *Regents of the University of Minnesota,* June 2016.

58 **don't need oxygen to form:** Woodward W. Fischer and Andrew H. Knoll, "An Iron Shuttle for Deepwater Silica in Late Archean and Early Paleoproterozoic Iron Formation," *Geological Society of America Bulletin* 121, no. 1–2 (2009): 222–35.

59 **some twenty plates:** "What Is Tectonic Shift?" US Department of Commerce, National Oceanic and Atmospheric Administration website.

60 **2,341 feet downward:** "Soudan Underground Mine Tours," Minnesota Department of Natural Resources website.

61 **three or four miles deeper underground:** Jim Essig, interview by the author, June 6, 2022.

62 **world's largest open-pit iron mine:** "Hull Rust Mine View—North Hibbing," City of Hibbing website.

62 **relocated beginning in 1919:** "About Hibbing," Hibbing Area Chamber of Commerce, https://www.hibbing.org/history/.

63 **still boasts 1,800 velvet seats:** Frank Edgerton Martin, "A Short History of Hibbing High School, an Iron Range Gem," *Enter,* August 31, 2017.

NOTES **251**

63 **More than 3 million feet of core:** "Drill Core Library," Minnesota Department of Natural Resources website.

65 **microscopically thin concentric layers:** Samuel Duncanson et al., "Reconstructing Diagenetic Mineral Reactions from Silicified Horizons of the Paleoproterozoic Biwabik Iron Formation, Minnesota," *American Mineralogist* 109, no. 2 (February 1, 2024): 339–58.

67 **the prestigious Crafoord Prize in Geoscience:** "The Crafoord Prize in Geosciences 2022," press release on Crafoord Prize website, January 29, 2022.

68 **availability of the nutrient phosphorus:** Andrew H. Knoll, "Food for Early Animal Evolution," *Nature* 548, no. 7669 (August 2017): 528–30.

70 **graph that illustrated the rise and fall:** Paul F. Hoffman and Daniel P. Schrag, "The Snowball Earth Hypothesis: Testing the Limits of Global Change," *Terra Nova* 14, no. 3 (June 2002): 146.

Five

77 **build up in protective mounds:** Laura M. Kehrl et al., "Glacimarine Sedimentation Processes at Kronebreen and Kongsvegen, Svalbard," *Journal of Glaciology* 57, no. 205 (2011): 841–47.

77 **tubes of sediments from the seafloor:** Laura Poppick, "Modern Depositional Processes Proximal to a Polythermal Tidewater Glacier Complex, Kronebreen-Kongsvegen, Kongsfjorden: Svalbard, Norway, 2010," Arctic Data Center, 2018.

79 **a scruffy-haired twenty-one-year-old:** "W. Brian Harland," Wikipedia, March 7, 2024.

80 **telltale signs of glacial activity:** W. Brian Harland, "Critical Evidence for a Great Infra-Cambrian Glaciation," *Geologische Rundschau* 54, no. 1 (May 1, 1964): 45–61.

80 **the most convincing evidence:** W. Brian Harland and Martin J. S. Rudwick, "The Great Infra-Cambrian Ice Age," *Scientific American* 211, no. 2 (1964): 28–37.

81 **yellow and gray carbonate rock:** Gabrielle Walker, *Snowball Earth: The Story of the Great Global Catastrophe That Spawned Life as We Know It* (New York: Crown Publishers, 2003), 67.

81 **mountainous slopes several miles above:** Paul F. Hoffman and Daniel P. Schrag, "Snowball Earth," *Scientific American* 282, no. 1 (January 1, 2000): 68–75.

81 **covered only about 8 percent of the planet:** "How Does Present Glacier Extent and Sea Level Compare to the Extent of Glaciers and Global Sea Level During the Last Glacial Maximum (LGM)?" US Geological Survey website.

81 **covered closer to 100 percent:** W. Brian Harland, "Origins and Assessment of Snowball Earth Hypotheses," *Geological Magazine* 144, no. 4 (July 2007): 633–42.

81 **dark and briny currents:** Paul Hoffman, email message to the author, February 10, 2019.

81 **howled around the equator:** Hoffman, email message, February 10, 2019.

81 **tiny ice crystals:** Hoffman, email message, February 10, 2019.

81 **swarmed alongside hydrothermal vents:** Anthonie Muller, "Animal Emergence During Snowball Earths by Thermosynthesis in Submarine Hydrothermal Vents," *Nature Precedings* 4 (June 17, 2009).

82 **follow-up studies of magnetic minerals:** Harland, "Origins and Assessment."

82 **"It is concluded":** Harland, "Critical Evidence," 46.

82 **Energized by these findings:** Walker, *Snowball Earth*, 71–74.

82 **don't drop uniformly:** Harland and Rudwick, "Great Infra-Cambrian Ice Age," 28–37.

82 **As early as 1891:** Harland and Rudwick, "Great Infra-Cambrian Ice Age," 28–37.

82 **the glacial interpretations of these rocks:** Harland and Rudwick, "Great Infra-Cambrian Ice Age," 28–37.

82 **measured between ten to one hundred meters thick:** Harland and Rudwick, "Great Infra-Cambrian Ice Age," 28–37.

83 **Earth was once much different:** Paul Hoffman, interview by the author, February 7, 2019.

83 **"Indeed," he and his coauthor, Martin Rudwick, lamented:** Harland and Rudwick, "Great Infra-Cambrian Ice Age," 33.

83 **"It can hardly be mere coincidence":** Harland and Rudwick, "Great Infra-Cambrian Ice Age," 34.

84 **growth and decay of Arctic ice:** Hoffman, interview, February 7, 2019.

85 **Budyko created a mathematical model:** M. I. Budyko, "The Effect of Solar Radiation Variations on the Climate of the Earth," *Tellus* 21, no. 5 (1969): 611–19.

85 **closer to 30 degrees latitude:** Daniel P. Schrag et al., "On the Initiation of a Snowball Earth," *Geochemistry, Geophysics, Geosystems* 3, no. 6 (2002): 1–21.

85 **wrote them off as implausible:** Hoffman, interview, February 7, 2019.

Six

86 **precursor to the theory of plate tectonics:** A. Hallam, "Alfred Wegener and the Hypothesis of Continental Drift," *Scientific American* 232, no. 2 (1975): 88–97.

87 **rarity in the geosciences:** "About Marie Tharp," Lamont-Doherty Earth Observatory website.

87 **sketched out those millions of data points:** D. Wright and H. Felt, "Marie Tharp: Discoverer of the Rift Valley of the Mid-Atlantic Ridge and Inventor of Marine Cartography," American Geophysical Union, Fall Meeting 2018, abstract #U22A-02.

87 **echo soundings collected:** Marie Tharp and Henry Frankel, "Mappers of the Deep: How Two Geologists Plotted the Mid-Atlantic Ridge and Made a Discovery That Revolutionized the Earth Sciences," *Natural History* (October 1, 1986): 1–6.

88 **argued for something along these lines:** Hallam, "Alfred Wegener," 88–97.

88 **dismissed by her male colleagues:** Neely Tucker, "Marie Tharp: Mapping the Ocean Floor," *Library of Congress Blogs*, August 9, 2021.

88 **1956 publication that shared:** Maurice Ewing and Bruce C. Heezen, "Some Problems of Antarctic Submarine Geology," In *Geophysical Monograph Series*, edited by A. P. Crary, L. M. Gould, E. O. Hulburt, Hugh Odishaw, and Waldo E. Smith (Washington, DC: American Geophysical Union, 2013), 75–81.

88 **"I worked in the background for most":** Sally Newcomb, "Marie Tharp's Discovery of the Mid Ocean Ridge Rift Valley in 1952," American Institute of Physics website, January 23, 2023.

NOTES

88 **a quarter of the postgraduate workforce:** Leila Gonzales, "Participation of Women in the Geoscience Profession," *American Geosciences Institute Geoscience Currents*, November 15, 2019.

88 **why women don't stay in the geosciences:** Ron E. Gray et al., "The Reasons Women Choose and Stay in a Geology Major: A Qualitative Multi-Case Analysis," *Innovation and Education* 3, no. 3 (December 2021): 1–14.

89 **"Women turned to science writing":** Anna Reser and Leila McNeill, *Forces of Nature: The Women Who Changed Science* (London: Frances Lincoln, 2021), 13.

89 **40,000 miles of valleys and ridges:** "Marie Tharp: Pioneering Mapmaker of the Ocean Floor," WHOI Women's Committee website.

89 **World Ocean Floor Map:** Heinrich C. Berann, Bruce C. Heezen, and Marie Tharp, "Manuscript painting of Heezen-Tharp 'World ocean floor' map," Library of Congress, 1977.

89 **When overlaid with seismograph readings:** Bruce C. Heezen, "The Rift in the Ocean Floor," *Scientific American* 203, no. 4 (October 1, 1960): 98–114.

90 **occasionally taking stomach-churning dips:** Paul Hoffman, interview by the author, December 17, 2021.

90 **Geological Survey began to shift its priorities:** Paul Hoffman, interview by the author, May 9, 2024.

91 **a paper that cleverly re-coined:** Joseph L. Kirschvink, "Late Proterozoic Low-Latitude Global Glaciation: The Snowball Earth," In: *The Proterozoic Biosphere*, edited by J. W. Schopf and C. Klein (Cambridge: Cambridge University Press, 1992), 51–52.

91 **a global ice age sounded "delicious" to Paul:** Paul Hoffman, email message to the author, March 31, 2022.

91 **the paradox they seemed to represent:** Paul Hoffman, email message to the author, May 29, 2024.

91 **seriously about Snowball Earth:** Hoffman, interview, May 9, 2024.

92 **preferentially pulls in carbon-12:** Paul F. Hoffman and Daniel P. Schrag, "Snowball Earth," *Scientific American* 282, no. 1 (January 1, 2000): 68–75.

92 **carbonate rocks without much carbon-12:** Hoffman and Schrag, "Snowball Earth," 68–75.

93 **looking for a life in geology:** Hoffman, email message, March 31, 2022.

93 **more than fifty months out in the field:** Hoffman, interview, February 7, 2019.

94 **most recent ice age surfaced in the 1800s:** W. H. Berger, "On the Discovery of the Ice Age: Science and Myth," *Geological Society, London, Special Publications* 273, no. 1 (January 2007): 271–78.

95 **Within thousands of years after:** Scott MacLennan et al., "The Arc of the Snowball: U-Pb Dates Constrain the Islay Anomaly and the Initiation of the Sturtian Glaciation," *Geology* 46, no. 6 (June 1, 2018): 539–42.

95 **powerful submarine currents:** Paul Hoffman, email message, February 10, 2019.

95 **energy bubbling out of these vents:** Anthonie Muller, "Animal Emergence During Snowball Earths by Thermosynthesis in Submarine Hydrothermal Vents," *Nature Precedings* 4 (June 17, 2009).

95 **dipping below negative 20 degrees Celsius:** Ryan C. Ewing et al., "New Constraints on Equatorial Temperatures During a Late Neoproterozoic Snowball Earth Glaciation," *Earth and Planetary Science Letters* 406 (November 2014): 110–22.

254   NOTES

95 **dust melted tiny holes in the ice:** Paul Hoffman et al., "Snowball Earth Climate Dynamics and Cryogenian Geology-Geobiology," *Science Advances* 3, no. 11
(November 3, 2017): e1600983.

95 **Snowball's most striking feature:** Paul Hoffman, interview, May 9, 2024.

95 **coalesced into hip-deep ponds:** Hoffman et al., "Snowball Earth Climate
Dynamics," e1600983.

96 **very earliest animals may have arisen:** Gordon D. Love and Roger E. Sum-
mons, "The Molecular Record of Cryogenian Sponges—A Response to Antcliffe
(2013)," *Palaeontology* 58, no. 6 (2015): 1131–36.

97 **Paul Hoffman agrees:** Paul Hoffman, interview by the author, February 26, 2019.

97 **Dan Schrag disagrees:** Daniel Schrag, interview by the author, April 6, 2022.

98 **a 2017 paper:** Hoffman et al., "Snowball Earth Climate Dynamics," e1600983.

98 **"Like geology itself":** Hoffman et al., "Snowball Earth Climate Dynamics," 4.

100 **Adam's research team had discovered:** Adam C. Maloof et al., "Possible
Animal-Body Fossils in Pre-Marinoan Limestones from South Australia," *Nature
Geoscience* 3, no. 9 (September 2010): 653–59.

101 **climate model published in 2000:** William T. Hyde et al., "Neoproterozoic
'Snowball Earth' Simulations with a Coupled Climate/Ice-Sheet Model," *Nature*
405, no. 6785 (May 2000): 425–29.

101 **Called the Waterbelt:** R. T. Pierrehumbert et al., "Climate of the Neoproterozoic," *Annual Review of Earth and Planetary Sciences* 39, no. 1 (May 30,
2011): 417–60.

104 **swivel and align with Earth's magnetic field:** Hoffman and Schrag, "Snowball
Earth," 68–75.

105 **pointing to origins near the equator:** W. Brian Harland and Martin J. S.
Rudwick, "The Great Infra-Cambrian Ice Age," *Scientific American* 211, no. 2
(1964): 28–37.

106 **arranged that way for millions of years:** Kai Lu et al., "Widespread Magmatic
Provinces at the Onset of the Sturtian Snowball Earth," *Earth and Planetary Science
Letters* 594 (September 2022): 117736.

106 **Volcanoes poked their peaks:** Kirschvink, "Late Proterozoic Low-Latitude
Global Glaciation," 51–52.

106 **as high as 660 times:** Paul M. Myrow, M. P. Lamb, and R. C. Ewing, "Rapid Sea
Level Rise in the Aftermath of a Neoproterozoic Snowball Earth," *Science* 360, no.
6389 (May 11, 2018): 649–51.

106 **a global average of 120 degrees Fahrenheit:** Myrow et al., "Rapid Sea Level
Rise," 2018.

106 **beyond 100 degrees Fahrenheit at the poles:** Hoffman et al., "Climate Dynamics and Cryogenian Geology-Geobiology," e1600983.

106 **over the course of thousands of years:** Myrow et al., "Rapid Sea Level
Rise," 2018.

107 **a few Australian geologists had identified:** For example, G. E. Williams, "Late
Neoproterozoic Periglacial Aeolian Sand Sheet, Stuart Shelf, South Australia," *Australian Journal of Earth Sciences* 45, no. 5 (October 1998): 733–41.

111 **I learned that Ryan hadn't received:** Ryan Ewing, interview by the author,
March 4, 2022.

114 **Ryan published his wedge findings:** Ewing et al., "New Constraints," 110–22.

NOTES

114 "Nothing hurries geology": Mark Twain, "Was the World Made for Man?" in *Letters from the Earth: Uncensored Writings*, edited by Bernard DeVoto (New York: First Perennial Classics, 2004), 221–26.

Seven

117 "This place . . . ," writes Annie Proulx: Annie Proulx, *The Shipping News* (New York: Touchstone, 1993), 32.

117 estimated 10,000 shipwrecks: "Divers Search for Wreck of Famous Schooner," *CBC News*, May 21, 2009.

119 widespread across at least eight paleocontinents: Judy P. Pu et al., "Dodging Snowballs: Geochronology of the Gaskiers Glaciation and the First Appearance of the Ediacaran Biota," *Geology* 44, no. 11 (November 2016): 955–58.

121 "the graveyard of the Atlantic": Mark King, email message to the author, May 29, 2024.

121 some do appear related to sponges: Shuhai Xiao, "Ediacaran Sponges, Animal Biomineralization, and Skeletal Reefs," *Proceedings of the National Academy of Sciences* 117, no. 35 (August 12, 2020): 20997–99.

121 cnidarians such as corals and jellyfish: F. S. Dunn et al., "A Crown-Group Cnidarian from the Ediacaran of Charnwood Forest, UK," *Nature Ecology & Evolution* 6, no. 8 (August 2022): 1095–1104.

121 Upward of 200 different types: Colin Barras, "These Half-Billion-Year-Old Creatures Were Animals—but Unlike Any Known Today," *Science*, August 8, 2018.

121 between about 579 million: Alexander G. Liu et al., "Effaced Preservation in the Ediacara Biota and Its Implications for the Early Macrofossil Record," *Palaeontology* 54, no. 3 (2011): 607–30.

121 to 538.8 million years old: Fred T. Bowyer et al., "Calibrating the Temporal and Spatial Dynamics of the Ediacaran–Cambrian Radiation of Animals," *Earth-Science Reviews* 225 (February 2022): 103913.

121 for the Ediacara Hills of Australia: Andrew H. Knoll et al., "The Ediacaran Period: A New Addition to the Geologic Time Scale," *Lethaia* 39, no. 1 (March 2006): 13–30.

121 the oldest and most extensive: "Mistaken Point," UNESCO World Heritage Centre website.

122 "the pizza discs": Liu et al., "Effaced Preservation," 607–30.

123 decomposed remains: Liu et al., "Effaced Preservation," 607–30.

124 "flowers in the rocks": "Mistaken Point Ecological Reserve and UNESCO World Heritage Site," Government of Newfoundland and Labrador website.

125 pancakes, ribbons, and threads: Stephen Jay Gould, *Wonderful Life: The Burgess Shale and the Nature of History* (New York: W. W. Norton & Company, 1990), 314.

127 "Since then," writes Stephen Jay Gould: Gould, *Wonderful Life*, 60.

127 most important transition in the entire: Luis A. Buatois, "*Treptichnus pedum* and the Ediacaran–Cambrian Boundary: Significance and Caveats," *Geological Magazine* 155, no. 1 (January 2018): 174–80.

127 best spot in the world to witness this transition: Martin Brasier, John Cowie,

NOTES

and Michael Taylor, "Decision on the Precambrian-Cambrian Boundary Strato-type," *Episodes Journal of International Geoscience* 17, no. 1 (1994): 3–8.

127 **440-meter-long stretch:** Brasier et al., "Decision on the Precambrian-Cambrian," 3–8.

127 **To be designated a golden spike:** "Global Boundary Stratotype Section and Points," International Commission on Stratigraphy website.

129 **Other *Treptichnus* fossils:** James G. Gehling et al., "Burrowing Below the Basal Cambrian GSSP, Fortune Head, Newfoundland," *Geological Magazine* 138, no. 2 (March 2001): 213–18.

Eight

135 **about 458 million years ago:** William J. McMahon and Neil S. Davies, "Evolution of Alluvial Mudrock Forced by Early Land Plants," *Science* 359, no. 6379 (March 2, 2018): 1022–24.

136 **descended from green algae:** Charles Francis Delwiche and Endymion Dante Cooper, "The Evolutionary Origin of a Terrestrial Flora," *Current Biology* 25, no. 19 (October 2015): R899–910.

136 **mosses and hornworts and liverworts:** Woodward W. Fischer, "Early Plants and the Rise of Mud," *Science* 359, no. 6379 (March 2, 2018): 994–95.

136 **what we'd recognize as forests:** Neil S. Davies, William J. McMahon, and Christopher M. Berry, "Earth's Earliest Forest: Fossilized Trees and Vegetation-Induced Sedimentary Structures from the Middle Devonian (Eifelian) Hangman Sandstone Formation, Somerset and Devon, SW England," *Journal of the Geological Society* 181, no. 4 (July 2024): jgs2023–204.

136 **ignited the very first wildfires:** Ian J. Glasspool and Robert A. Gastaldo, "Silurian Wildfire Proxies and Atmospheric Oxygen," *Geology* 50, no. 9 (September 1, 2022): 1048–52.

136 **animals that grew ever larger:** Tais W. Dahl and Susanne K. M. Arens, "The Impacts of Land Plant Evolution on Earth's Climate and Oxygenation State—An Interdisciplinary Review," *Chemical Geology* 547 (August 2020): 119665.

136 **better swimmers and better predators:** Dahl and Arens, "Impacts of Land Plant Evolution," 119665.

136 **better at having sex:** Andrew M. Bush, Gene Hunt, and Richard K. Bambach, "Sex and the Shifting Biodiversity Dynamics of Marine Animals in Deep Time," *Proceedings of the National Academy of Sciences* 113, no. 49 (December 6, 2016): 14073–78.

136 **Scottish geologist who discovered the species:** "Robert Dick," Wikipedia website, June 23, 2023.

136 **The fish's sex organ:** John A. Long et al., "Copulation in Antiarch Placoderms and the Origin of Gnathostome Internal Fertilization," *Nature* 517, no. 7533 (January 2015): 196–99.

136 **it changed the shape of topography:** Martin R. Gibling and Neil S. Davies, "Palaeozoic Landscapes Shaped by Plant Evolution," *Nature Geoscience* 5, no. 2 (February 2012): 99–105.

137 **sinuous channels such as parts of the Amazon:** José Antonio Constantine et

NOTES 257

al., "Sediment Supply as a Driver of River Meandering and Floodplain Evolution in the Amazon Basin," *Nature Geoscience* 7, no. 12 (December 2014): 899–903.

137 **Mississippi Rivers today:** "Meandering Mississippi River," NASA Earth Observatory, July 19, 2020.

137 **attracts organic detritus:** Jordon D. Hemingway et al., "Mineral Protection Regulates Long-Term Global Preservation of Natural Organic Carbon," *Nature* 570, no. 7760 (June 2019): 228–31.

137 **20 percent of human carbon dioxide emissions:** Judy Q. Yang et al., "4D Imaging Reveals Mechanisms of Clay-Carbon Protection and Release," *Nature Communications* 12, no. 1 (January 27, 2021): 622.

138 **"I want you to fill your hands with mud":** Mary Oliver, "Rice," in *New and Selected Poems: Volume One* (Boston: Beacon Press, 2005), 38.

138 **"Without mud, there can be no lotus":** Thich Nhat Hanh, *No Mud, No Lotus: The Art of Transforming Suffering* (Berkeley, CA: Parallax Press, 2014), 13.

139 **sometime during the Middle Ordovician:** Neil Davies, interview by the author, March 19, 2020.

139 **majority of the scientific community:** Davies, interview, March 19, 2020.

139 **collection of early fish fossils:** Greg Graffin, "A New Locality of Fossiliferous Harding Sandstone: Evidence for Freshwater Ordovician Vertebrates," *Journal of Vertebrate Paleontology* 12, no. 1 (March 6, 1992): 1–10.

139 **others disagreed with his interpretation:** Neil S. Davies et al., "Ichnology, Palaeoecology and Taphonomy of a Gondwanan Early Vertebrate Habitat: Insights from the Ordovician Anzaldo Formation, Bolivia," *Palaeogeography, Palaeoclimatology, Palaeoecology* 249, no. 1–2 (June 2007): 18–35.

140 **Graffin's theory was already unpopular:** Davies, interview, March 19, 2020.

141 **As early as the 1940s:** Thorolf Vogt and Gunnar Horn, "Geology of a Middle Devonian Cannel Coal from Spitsbergen," *Norwegian Journal of Geology* 21 (1941): 1–12.

141 **all 704 known river deposits:** McMahon and Davies, "Evolution of Alluvial Mudrock," 1022–24.

141 **mudrock increased by more than tenfold:** McMahon and Davies, "Evolution of Alluvial Mudrock," 1022–24.

141 **fine-grained deposits rose:** McMahon and Davies, "Evolution of Alluvial Mudrock," 1022–24.

142 **erosive process called chemical weathering:** Eugene F. Kelly, Oliver A. Chadwick, and Thomas E. Hilinski, "The Effect of Plants on Mineral Weathering," *Biogeochemistry* 42, no. 1/2 (1998): 21–53.

Nine

144 **land plants grew widespread:** Martin R. Gibling and Neil S. Davies, "Palaeozoic Landscapes Shaped by Plant Evolution," *Nature Geoscience* 5, no. 2 (February 2012): 99–105.

147 **"no vestige of a beginning":** James Hutton, "Theory of the Earth; or an Investigation of the Laws Observable in the Composition, Dissolution, and Restoration of Land upon the Globe," *Transactions of the Royal Society of Edinburgh* 1 (1788): 96.

258 NOTES

147 **world's largest millipede fossil:** Neil S. Davies et al., "The Largest Arthropod in Earth History: Insights from Newly Discovered *Arthropleura* Remains (Serpukhovian Stainmore Formation, Northumberland, England)," *Journal of the Geological Society* 179, no. 3 (May 2022): jgs2021–115.

150 **Carbon dioxide arose in the atmosphere:** Michael P. D'Antonio, Daniel E. Ibarra, and C. Kevin Boyce, "Land Plant Evolution Decreased, Rather Than Increased, Weathering Rates," *Geology* 48, no. 1 (January 1, 2020): 29–33.

150 **24,000 pounds:** "Prairies and Our Lakes," Clean Lakes Alliance website.

151 **"Sex That Moves Mountains":** Alexander K. Fremier, Brian J. Yanites, and Elowyn M. Yager, "Sex That Moves Mountains: The Influence of Spawning Fish on River Profiles over Geologic Timescales," *Geomorphology* 305 (March 2018): 163–72.

Ten

152 **bestselling book on this notion:** Jeremy Megraw, "The Importance of Earthworms: Darwin's Last Manuscript," New York Public Library website, April 19, 2012.

152 **treatise on earthworms:** Charles R. Darwin, *The Formation of Vegetable Mould, Through the Action of Worms, with Observations on Their Habits* (London: John Murray, 1881).

152 **as much as ten tons of soil:** Darwin, *Formation of Vegetable Mould*, 258.

152 **"The result for a country the size":** Darwin, *Formation of Vegetable Mould*, 258.

153 **not so much a binary system:** Michal Tal, interview by the author, March 20, 2023.

153 **that braided channel will naturally:** Tal, interview, March 20, 2023.

154 **North America's tallest:** "Whooping Crane," Platte River Recovery Implementation Program website.

154 **they would sleep atop sandbars:** Tal, interview, March 20, 2023.

156 **already begun buying up miles of land:** Tal, interview, March 20, 2023.

156 **garner support for restoration practices:** Tal, interview, March 20, 2023.

156 **validating their observations:** Martin R. Gibling and Neil S. Davies, "Palaeozoic Landscapes Shaped by Plant Evolution," *Nature Geoscience* 5, no. 2 (February 2012): 99–105.

156 **The present is the key to the past:** This concept of uniformitarianism is described throughout: Charles Lyell, *Principles of Geology: Being an Attempt to Explain the Former Changes of the Earth's Surface, by Reference to Causes Now in Operation*, vol. 1 (London: John Murray, 1830).

157 **a geologic field guide:** Pat Meere, Ivor MacCarthy, Reginald Reavy, Alistair Allen, Ken Higgs, *Geology of Ireland: A Field Guide* (Doughcloyne, Wilton, Cork, UK: The Collins Press, 2013).

157 **the possibility of trace fossils:** Meere et al., *Geology of Ireland*, 140.

160 **findings on the rise of mud:** William J. McMahon and Neil S. Davies, "Evolution of Alluvial Mudrock Forced by Early Land Plants," *Science* 359, no. 6379 (March 2, 2018): 1022–24.

160 **including . . . *Dollyphyton boucotii*:** Kristin Strommer, "Geologist Helps Confirm Date of Earliest Land Plants on Earth," University of Oregon website, November 3, 2020.

NOTES

161 **he dabbles in other events as well:** Woodward W. Fischer, "Early Plants and the Rise of Mud," *Science* 359, no. 6379 (March 2, 2018): 994–95.

161 **trapping organic material:** Judy Q. Yang et al., "4D Imaging Reveals Mechanisms of Clay-Carbon Protection and Release," *Nature Communications* 12, no. 1 (January 27, 2021): 622.

162 **before long, he took a job elsewhere:** Sarah Zeichner, interview by the author, April 18, 2023.

163 **paper that she copublished in *Science*:** Sarah S. Zeichner et al., "Early Plant Organics Increased Global Terrestrial Mud Deposition Through Enhanced Flocculation," *Science* 371, no. 6528 (January 29, 2021): 526–29.

165 **exude cocktails of sugars and acids:** Eugene F. Kelly, Oliver A. Chadwick, and Thomas E. Hilinski, "The Effect of Plants on Mineral Weathering," *Biogeochemistry* 42, no. 1/2 (1998): 21–53.

165 **band of white chalky material:** Ana M. Alonso-Zarza, "Palaeoenvironmental Significance of Palustrine Carbonates and Calcretes in the Geological Record," *Earth-Science Reviews* 60, no. 3–4 (February 2003): 261–98.

165 **ferricretes develop in soils:** Lawrence H. Tanner and Mohamed A. Khalifa, "Origin of Ferricretes in Fluvial-Marine Deposits of the Lower Cenomanian Bahariya Formation, Bahariya Oasis, Western Desert, Egypt," *Journal of African Earth Sciences* 56, nos. 4–5 (March 2010): 179–89.

Eleven

167 **Hutton spent the years following:** Jack Repcheck, *The Man Who Found Time: James Hutton and the Discovery of the Earth's Antiquity* (Reading, MA: Perseus, 2003), 18.

167 **could no longer easily explore:** Repcheck, *Man Who Found Time*, 19.

167 **James Hall, who offered up a boat:** Repcheck, *Man Who Found Time*, 16.

169 **"For Hutton," writes biographer Jack Repcheck:** Repcheck, *Man Who Found Time*, 18.

169 **Hutton narrated the scene:** Repcheck, *Man Who Found Time*, 13–24.

169 **"On us who saw these phenomena":** John Playfair, *Biographical Account of James Hutton, M.D. F.R.S. Ed.* (United Kingdom: n.p., 1797), 34–35.

170 **red sandstone that topped it all off:** Colin MacFadyen, Martina Kölbl-Ebert, and Helen Fallas, "Siccar Point Hutton's Unconformity," IUGS International Commission on Geoheritage website, 2024.

170 **parking lot just beyond Wine Strand:** Pat Meere, Ivor MacCarthy, Reginald Reavy, Alistair Allen, Ken Higgs, *Geology of Ireland: A Field Guide* (Doughcloyne, Wilton, Cork, UK: The Collins Press, 2013), 141–43.

170 **cliffs made of a similar red sandstone:** R. S. Kendall, "The Old Red Sandstone of Britain and Ireland—A Review," *Proceedings of the Geologists' Association* 128, no. 3 (June 2017): 409–21.

171 **to borrow Playfair's phrase:** Playfair, *Biographical Account*, 34–35.

Twelve

173 **this collection of trees:** William E. Stein et al., "Mid-Devonian *Archaeopteris* Roots Signal Revolutionary Change in Earliest Fossil Forests," *Current Biology* 30, no. 3 (February 2020): 421–31.e2.

173 **evidence of a slightly older forest:** Neil S. Davies, William J. McMahon, and Christopher M. Berry, "Earth's Earliest Forest: Fossilized Trees and Vegetation-Induced Sedimentary Structures from the Middle Devonian (Eifelian) Hangman Sandstone Formation, Somerset and Devon, SW England," *Journal of the Geological Society* 181, no. 4 (July 2024): jgs2023-204.

174 **newly forming logjams:** Martin R. Gibling and Neil S. Davies, "Palaeozoic Landscapes Shaped by Plant Evolution," *Nature Geoscience* 5, no. 2 (February 2012): 99–105.

174 **massive marine algae blooms:** R. S. Kendall, "The Old Red Sandstone of Britain and Ireland—A Review," *Proceedings of the Geologists' Association* 128, no. 3 (June 2017): 409–21.

174 **up to 80 percent of all shallow marine life:** Kendall, Old Red Sandstone," 409–21.

174 **"Mud is providing a totally different medium":** Anthony Shillito, interview by the author, February 5, 2020.

174 **burrowing through coarse material:** Shillito, interview, February 5, 2020.

175 **This loosening of silts and clays:** Lidya Tarhan, interview by the author, June 30, 2020.

175 **"They are phenomenal engineers":** Tarhan, interview, June 30, 2020.

175 **"The landscapes and seascapes":** Neil S. Davies et al., "Evolutionary Synchrony of Earth's Biosphere and Sedimentary-Stratigraphic Record," *Earth-Science Reviews* 201 (February 2020): 102979.

177 **a role in the global cooling:** Bo Chen et al., "Devonian Paleoclimate and Its Drivers: A Reassessment Based on a New Conodont $\delta 18O$ Record from South China," *Earth-Science Reviews* 222 (November 1, 2021): 103814.

178 **It would either free-fall:** Michael D'Antonio, interview by the author, May 30, 2023.

179 **Based on computer models:** Michael P. D'Antonio, Daniel E. Ibarra, and C. Kevin Boyce, "Land Plant Evolution Decreased, Rather Than Increased, Weathering Rates," *Geology* 48, no. 1 (January 1, 2020): 29–33.

180 **a tattered manuscript:** Geologists use this analogy to sum up a passage in Charles Darwin, *On the Origin of Species by Means of Natural Selection, or Preservation of Favoured Races in the Struggle for Life* (London: John Murray, 1859), 310–11.

180 **a paper he coauthored in 2021:** Neil S. Davies and Anthony P. Shillito, "True Substrates: The Exceptional Resolution and Unexceptional Preservation of Deep Time Snapshots on Bedding Surfaces," *Sedimentology* 68, no. 7 (December 2021): 3307–56.

181 **"As eerily familiar snapshots":** Davies and Shillito, "True Substrates," 3351.

NOTES   **261**

Thirteen

187 **propelled by the formation of sea ice:** US Department of Commerce, National Oceanic and Atmospheric Administration, "Thermohaline Circulation," NOAA's National Ocean Service Education website.

187 **sites of deepwater formation:** US Department of Commerce, National Oceanic and Atmospheric Administration, "The Global Conveyor Belt," NOAA's National Ocean Service Education website.

187 **once every twenty seconds:** Kim Ann Zimmermann, "The Circulatory System: An Amazing Circuit That Keeps Our Bodies Going," *Live Science*, August 9, 2022.

187 **once every 1,000 years:** "Global Conveyor Belt," NOAA website.

188 **35 percent during the Carboniferous period:** Uwe Brand et al., "Atmospheric Oxygen of the Paleozoic," *Earth-Science Reviews* 216 (May 1, 2021): 103560.

188 **dragonfly-like insects swelled larger than seagulls:** Haichun Zhang et al., "The Largest Known Odonate in China: Hsiufua Chaoi Zhang et Wang, gen. et sp. nov. from the Middle Jurassic of Inner Mongolia," *Chinese Science Bulletin* 58, no. 13 (May 2013): 1579–84.

188 **scorpions lurched longer than woodchucks:** Andrew J. Jeram, "Scorpions from the Viséan of East Kirkton, West Lothian, Scotland, with a Revision of the Infraorder Mesoscorpionina," *Earth and Environmental Science Transactions of the Royal Society of Edinburgh* 84, no. 3–4 (1993): 283–99.

188 **scuttled larger than adult humans:** Neil S. Davies et al., "The Largest Arthropod in Earth History: Insights from Newly Discovered *Arthropleura* Remains (Serpukhovian Stainmore Formation, Northumberland, England)," *Journal of the Geological Society* 179, no. 3 (May 2022): jgs2021–115.

188 **single massive supercontinent, Pangea:** "What Was Pangea?" US Geological Survey website.

188 **Metamorphosing insects:** James W. Truman and Lynn M. Riddiford, "The Evolution of Insect Metamorphosis: A Developmental and Endocrine View," *Philosophical Transactions of the Royal Society B: Biological Sciences* 374, no. 1783 (October 14, 2019): 20190070.

188 **reptiles emerged for the first time:** Lorenzo Marchetti et al., "Tracking the Origin and Early Evolution of Reptiles," *Frontiers in Ecology and Evolution* 9 (July 1, 2021): 696511.

188 **subterranean pipes and out of craters:** Henrik H. Svensen et al., "Sills and Gas Generation in the Siberian Traps," *Philosophical Transactions of the Royal Society A: Mathematical, Physical and Engineering Sciences* 376, no. 2130 (October 13, 2018): 20170080.

189 **next tens of thousands of years:** Seth D. Burgess, Samuel Bowring, and Shuzhong Shen, "High-Precision Timeline for Earth's Most Severe Extinction," *Proceedings of the National Academy of Sciences* 111, no. 9 (March 4, 2014): 3316–21.

189 **5 million square kilometers:** L. E. Augland et al., "The Main Pulse of the Siberian Traps Expanded in Size and Composition," *Scientific Reports* 9, no. 1 (December 10, 2019): 18723.

189 **lava . . . one kilometer thick:** Jennifer Chu, "Siberian Traps Likely Culprit for End-Permian Extinction," *MIT News*, September 16, 2015.

NOTES

189 **10 degrees Celsius and sea ice vanished:** Michael J. Benton, "Hyperthermal-Driven Mass Extinctions: Killing Models During the Permian–Triassic Mass Extinction," *Philosophical Transactions of the Royal Society A: Mathematical, Physical and Engineering Sciences* 376, no. 2130 (October 13, 2018): 20170076.

189 **96 percent of all marine . . . 70 percent of terrestrial:** Justin L. Penn et al., "Temperature-Dependent Hypoxia Explains Biogeography and Severity of End-Permian Marine Mass Extinction," *Science* 362, no. 6419 (December 7, 2018): eaat1327.

189 **10 million square kilometers of magma:** J.H.F.L. Davies et al., "End-Triassic Mass Extinction Started by Intrusive CAMP Activity," *Nature Communications* 8, no. 1 (May 31, 2017): 15596.

189 **four pulses that spanned some 600,000 years:** Terrence J. Blackburn et al., "Zircon U-Pb Geochronology Links the End-Triassic Extinction with the Central Atlantic Magmatic Province," *Science* 340, no. 6135 (May 24, 2013): 941–45.

189 **76 percent of marine and terrestrial species:** Jennifer Chu, "Huge and Widespread Volcanic Eruptions Triggered the End-Triassic Extinction," *MIT News,* March 21, 2013.

189 **"Every pulse of volcanic activity":** Jennifer C. McElwain, Marlene Hill Donnelly, and Ian J. Glasspool, *Tropical Arctic: Lost Plants, Future Climates, and the Discovery of Ancient Greenland* (Chicago: University of Chicago Press, 2021), 81.

190 **"There is always life":** McElwain et al., *Tropical Arctic,* 67.

190 **2,500 human generations:** McElwain et al., *Tropical Arctic,* 85–86.

190 **butanes and benzenes and ozone–depleting gases:** Henrik Svensen et al., "Siberian Gas Venting and the End-Permian Environmental Crisis," *Earth and Planetary Science Letters* 277, no. 3–4 (January 2009): 490–500.

190 **ten to one hundred times faster:** McElwain et al., *Tropical Arctic,* 62.

192 **have no insulating powers:** Sarah Fecht, "How Exactly Does Carbon Dioxide Cause Global Warming?" *State of the Planet,* February 25, 2021.

192 **placental mammals begin to rise:** Maureen A. O'Leary et al., "The Placental Mammal Ancestor and the Post–K-Pg Radiation of Placentals," *Science* 339, no. 6120 (February 8, 2013): 662–67.

192 **between 14 to 25 degrees Celsius warmer:** Jan Landwehrs et al., "Investigating Mesozoic Climate Trends and Sensitivities with a Large Ensemble of Climate Model Simulations," *Paleoceanography and Paleoclimatology* 36, no. 6 (June 2021): e2020PA004134.

193 **About 70 percent of all modern oil deposits:** "Oil Formation," Energy Education website.

193 **fossilized remains of marine plankton:** "Oil Formation," Energy Education website.

193 **Manhattan joined:** Ed Shanahan, "New York City's Air Was 'Very Unhealthy,' the Mayor Said," *The New York Times,* June 7, 2023.

193 **Chicago . . . lowest-ranking air quality:** Jenna Smith and Nell Salzman, "Chicago's Air Quality Worst in the World, Global Index Shows," *Chicago Tribune,* June 29, 2023.

194 **to some 200,000 evacuations:** David Ljunggren, "Canada Wildfire: All 20,000 Yellowknife Residents Evacuating," Reuters, August 18, 2023.

NOTES

194 **tens of thousands of Indigenous people:** Brent McDonald, Matt Joycey, and Ben Laffin, "Canada Is Ravaged by Fire. No One Has Paid More Dearly Than Indigenous People," *The New York Times*, July 29, 2023.

194 **2,000 acres of crops . . . $15 million in losses:** Colin A. Young, "11 Months After Flooding Ruined Berkshire Crops, No Farms Have Been Lost, the State's Director of Rural Affairs Says," *The Berkshire Eagle*, June 17, 2024.

194 **99 percent of the rest of the ocean:** Andrew J. Pershing et al., "Slow Adaptation in the Face of Rapid Warming Leads to Collapse of the Gulf of Maine Cod Fishery," *Science* 350, no. 6262 (November 13, 2015): 809–12.

194 **top five hottest summers:** "Gulf of Maine Warming Update: Summer 2023," Gulf of Maine Research Institute website, October 3, 2023.

194 **Shrimp . . . have declined:** "Northern Shrimp Population Collapse Linked to Warming Ocean Temperatures, Squid Predation," NOAA Fisheries website, September 28, 2021.

194 **cod populations have declined:** Pershing et al., "Slow Adaptation in the Face of Rapid Warming."

194 **phytoplankton . . . have grown less productive:** William M. Balch et al., "Changing Hydrographic, Biogeochemical, and Acidification Properties in the Gulf of Maine as Measured by the Gulf of Maine North Atlantic Time Series, GNATS, Between 1998 and 2018," *Journal of Geophysical Research: Biogeosciences* 127, no. 6 (June 2022): e2022JG006790.

194 **longfin squid:** R. Anne Richards and Margaret Hunter, "Northern Shrimp *Pandalus borealis* Population Collapse Linked to Climate-Driven Shifts in Predator Distribution," *PLOS ONE* 16, no. 7 (July 21, 2021): e0253914.

194 **pulled two seahorses from his traps:** Elaine Jones, "Seahorses in Boothbay Harbor?" *Boothbay Register*, September 17, 2018.

195 **oldest evidence of wildfire:** Ian J. Glasspool and Robert A. Gastaldo, "Silurian Wildfire Proxies and Atmospheric Oxygen," *Geology* 50, no. 9 (September 1, 2022): 1048–52.

195 **fivefold increase in charcoal:** Claire M. Belcher et al., "Increased Fire Activity at the Triassic/Jurassic Boundary in Greenland Due to Climate-Driven Floral Change," *Nature Geoscience* 3, no. 6 (June 2010): 426–29.

196 **420-million-year-old deposits:** I. J. Glasspool, D. Edwards, and L. Axe, "Charcoal in the Silurian as Evidence for the Earliest Wildfire," *Geology* 32, no. 5 (2004): 381.

196 **another 10 million years:** Glasspool and Gastaldo, "Silurian Wildfire Proxies."

197 **more than twenty feet high:** Erik A. Hobbie and C. Kevin Boyce, "Carbon Sources for the Palaeozoic Giant Fungus *Prototaxites* Inferred from Modern Analogues," *Proceedings of the Royal Society B: Biological Sciences* 277, no. 1691 (July 22, 2010): 2149–56.

198 **oxygen levels below 16 percent:** Glasspool and Gastaldo, "Silurian Wildfire Proxies."

198 **oxygen concentrations had grown greater:** Benjamin J. W. Mills et al., "Evolution of Atmospheric O_2 Through the Phanerozoic, Revisited," *Annual Review of Earth and Planetary Sciences* 51, no. 1 (May 31, 2023): 253–76.

198 **Just 1 degree Celsius of warming:** N. Reeve and R. Toumi, "Lightning Activity

NOTES

as an Indicator of Climate Change," *Quarterly Journal of the Royal Meteorological Society* 125, no. 555 (April 1999): 893–903.

198 **shape of leaves changed significantly:** Belcher et al., "Increased Fire Activity."

199 **phosphorus could have spilled into the ocean:** Glasspool and Gastaldo, "Silurian Wildfire Proxies."

201 **Thousands of these little mouths:** Jared Dashoff, "Scientists Discover Mechanism Plants Use to Control 'Mouths,'" *Science Matters*, US National Science Foundation website, December 7, 2022.

201 **quadrupling of carbon dioxide levels:** Sarah J. Baker et al., "CO_2-Induced Biochemical Changes in Leaf Volatiles Decreased Fire-Intensity in the Run-Up to the Triassic–Jurassic Boundary," *New Phytologist* 235, no. 4 (August 2022): 1442–54.

201 **3 to 4 degree Celsius increase:** Baker et al., "CO_2-Induced Biochemical Changes."

201 **"For some educational purposes":** Sophie H. Eckerson, "The Number and Size of the Stomata," *Botanical Gazette* 46, no. 3 (September 1908): 221.

201 **"Probably because of incipient wilting":** Eckerson, "Number and Size of the Stomata," 224.

202 **40 percent decline:** F. I. Woodward, "Stomatal Numbers Are Sensitive to Increases in CO_2 from Pre-Industrial Levels," *Nature* 327, no. 6123 (June 1987): 617–18.

202 **carbon dioxide levels increased some 21 percent:** Woodward, "Stomatal Numbers," 618.

202 **carbon dioxide levels based on the density of stomata:** Jennifer C. McElwain and William G. Chaloner, "Stomatal Density and Index of Fossil Plants Track Atmospheric Carbon Dioxide in the Palaeozoic," *Annals of Botany* 76, no. 4 (1995): 389–95.

202 **global temperature rise of 2 degrees Celsius:** "Global and European Temperatures," European Environment Agency website, June 26, 2024.

202 **physicist Robert Hooke correctly assumed:** R. Hooke, *The Posthumous Works of Robert Hooke, . . . Containing His Cutlerian Lectures, and Other Discourses, Read at the Meetings of the Illustrious Royal Society. . . . Illustrated with Sculptures. To These Discourses Is Prefixt the Author's Life, . . . Publish'd by Richard Waller* (London: Sam. Smith and Benj. Walford, 1705), 342–43.

203 **alligators and tortoises in the Canadian High Arctic:** Jaelyn J. Eberle et al., "Seasonal Variability in Arctic Temperatures During Early Eocene Time," *Earth and Planetary Science Letters* 296, no. 3–4 (August 2010): 481–86.

203 **advocate for using stratigraphic data:** Jessica E. Tierney et al., "Past Climates Inform Our Future," *Science* 370, no. 6517 (November 6, 2020): eaay3701.

203 **proxies of choice include fossilized fats:** Jessica E. Tierney and Martin P. Tingley, "A TEX86 Surface Sediment Database and Extended Bayesian Calibration," *Scientific Data* 2, no. 1 (June 23, 2015): 150029.

204 **temperatures beyond 100 degrees Fahrenheit:** Jessica Tierney, interview by the author, December 8, 2022.

204 **can more confidently turn to those models:** Tierney, interview, December 8, 2022.

205 **shoveled elephant dung:** Karen Chin, interview by the author, December 19, 2022.

NOTES

206 **calibrations that Jessica Tierney . . . had published:** Tierney and Tingley, "TEX86 Surface Sediment Database."

207 **as high 69 degrees Fahrenheit:** James R. Super et al., "Late Cretaceous Climate in the Canadian Arctic: Multi-Proxy Constraints from Devon Island," *Palaeogeography, Palaeoclimatology, Palaeocology* 504 (September 2018): 1–22.

207 **marine food chains . . . relatively short:** Karen Chin et al., "Life in a Temperate Polar Sea: A Unique Taphonomic Window on the Structure of a Late Cretaceous Arctic Marine Ecosystem," *Proceedings of the Royal Society B: Biological Sciences* 275, no. 1652 (December 7, 2008): 2675–85.

207 **10 degrees Celsius over . . . 60,000 years:** Seth D. Burgess et al., "High-Precision Timeline," *Proceedings of the National Academy of Sciences* 111, no. 9 (March 4, 2014): 2014.

207 **4 degrees Celsius over just 200 years:** Intergovernmental Panel on Climate Change, "Regional Fact Sheet—Ocean," *Sixth Assessment Report*, Working Group I—The Physical Science Basis, 2021.

208 **40 percent of amphibian species:** Jennifer A. Luedtke et al., "Ongoing Declines for the World's Amphibians in the Face of Emerging Threats," *Nature* 622, no. 7982 (October 12, 2023): 308–14.

208 **species can adapt and evolve on human timescales:** Agnes Holstad et al., "Evolvability Predicts Macroevolution Under Fluctuating Selection," *Science* 384, no. 6696 (May 10, 2024): 688–93.

Fourteen

209 **ice-free and, in some places, thickly forested:** Paul Olsen et al., "Arctic Ice and the Ecological Rise of the Dinosaurs," *Science Advances* 8, no. 26 (July 2022): eabo6342.

209 **birth of the modern Rocky Mountains:** "Geology of Rocky Mountain National Park," US Geological Survey website.

209 **dismantling pieces of Pangea:** Sabin Zahirovic et al., "Tectonic Evolution and Deep Mantle Structure of the Eastern Tethys Since the Latest Jurassic," *Earth-Science Reviews* 162 (November 1, 2016): 293–337.

209 **oceans had temporarily acidified:** Calum P. Fox et al., "Two-Pronged Kill Mechanism at the End-Triassic Mass Extinction," *Geology* 50, no. 4 (January 5, 2022): 448–53.

209 **much of their dissolved oxygen:** Tianchen He et al., "Shallow Ocean Oxygen Decline During the End-Triassic Mass Extinction," *Global and Planetary Change* 210 (March 1, 2022): 103770.

210 **world's leading stegosaur scientist:** "Hodson Award: Dr. Susannah C. R. Maidment," *Palaeontology Newsletter*, no. 94 (March 2017): 11.

212 **Since the 1870s:** "The Morrison Formation—Fossils and Paleontology," US National Park Service website, July 8, 2022.

212 **millions of dollars in funding:** "Lilly Endowment Awards $9 Million for Fossil Excavation Project," *Philanthropy News Digest*, March 18, 2019.

212 **156 million and 147 million years old:** Susannah C. R. Maidment and Adrian Muxworthy, "A Chronostratigraphic Framework for the Upper Jurassic Morrison

266

NOTES

Formation, Western U.S.A.," *Journal of Sedimentary Research* 89, no. 10 (October 29, 2019): 1017–38.

212 **comprehensive chronology:** Maidment and Muxworthy, "Chronostratigraphic Framework."

213 **245 additional ones:** Maidment and Muxworthy, "Chronostratigraphic Framework."

213 **peaks around the humid tropics:** James H. Brown, "Why Are There So Many Species in the Tropics?" *Journal of Biogeography* 41, no. 1 (January 2014): 8–22.

214 **more arid and less biologically productive:** Maidment and Muxworthy, "Chronostratigraphic Framework."

215 **fossil diggers had found these bones:** Susie Maidment, interview by the author, June 24, 2019.

215 **twenty-year lease on the land:** "A Fall Haul of Fossil Treasures Is Making Its Way to the Children's Museum of Indianapolis from the Jurassic Mile," Children's Museum of Indianapolis website.

217 **most complete stegosaur skeleton:** Susannah Catherine Rose Maidment et al., "The Postcranial Skeleton of an Exceptionally Complete Individual of the Plated Dinosaur *Stegosaurus stenops* (Dinosauria: Thyreophora) from the Upper Jurassic Morrison Formation of Wyoming, U.S.A.," *PLOS ONE* 10, no. 10 (October 14, 2015): e0138352.

218 **meandering streams and muddy floodplains:** Maidment and Muxworthy, "Chronostratigraphic Framework."

222 **something like one ton:** John Pickrell, "Sauropods Grew Big by Munching 'Superfoods' with Sturdy Beaks," *Science*, October 17, 2019.

229 **compiled a chronology similar:** David A. Eberth et al., "Calibrating Geologic Strata, Dinosaurs, and Other Fossils at Dinosaur Provincial Park (Alberta, Canada) Using a New CA-ID-TIMS U–Pb Geochronology," *Canadian Journal of Earth Sciences* 60, no. 12 (December 1, 2023): 1627–46.

229 **crest shapes in hadrosaurs:** Susie Maidment, interview by the author, June 24, 2019.

231 **iridium in concentrations orders of magnitude greater:** Steven Goderis et al., "Globally Distributed Iridium Layer Preserved Within the Chicxulub Impact Structure," *Science Advances* 7, no. 9 (February 26, 2021): eabe3647.

231 **more than one hundred miles:** Alfio Alessandro Chiarenza et al., "Asteroid Impact, Not Volcanism, Caused the End-Cretaceous Dinosaur Extinction," *Proceedings of the National Academy of Sciences* 117, no. 29 (July 21, 2020): 17084–93.

231 **tropics may have hovered around freezing:** Chiarenza et al., "Asteroid Impact."

231 **Three-quarters of all animal species:** Emily Osterloff, "How an Asteroid Ended the Age of the Dinosaurs," Natural History Museum, London website.

231 **on the order of tens of thousands of years:** Paul R. Renne et al., "Time Scales of Critical Events Around the Cretaceous-Paleogene Boundary," *Science* 339, no. 6120 (February 8, 2013): 684–87.

231 **within nine months of the impact:** Mindy Weisberger, "Darkness Caused by Dino-Killing Asteroid Snuffed Out Life on Earth in 9 Months," *Live Science*, December 22, 2021.

NOTES 267

Epilogue: Us

236 **established beyond question:** Rachel Carson, *The Sea Around Us* (New York: Oxford University Press, 1951), 168.

236 **"The frigid top of the world":** Carson, *Sea Around Us*, 168.

236 **sold some 250,000 copies:** "Legacy of Rachel Carson's *Silent Spring*," American Chemical Society National Historic Chemical Landmarks website.

236 **bestseller list for eighty-six weeks:** "Rachel Carson," National Women's Hall of Fame website.

236 **more than 1 million copies . . . twenty-eight languages:** "The Sea Around Us," National Book Foundation website.

237 **hottest year on record:** "2023 Was the Warmest Year in the Modern Temperature Record," NOAA National Centers for Environmental Information, January 17, 2024.

237 **One 2023 study:** Boen Zhang et al., "Higher Exposure of Poorer People to Emerging Weather Whiplash in a Warmer World," *Geophysical Research Letters* 50, no. 21 (November 16, 2023): e2023GL105640.

237 **Carson wrote of:** The following paragraph pulls details from *The Sea Around Us*, 168–71.

237 **"The evidence that the top of the world":** Carson, *Sea Around Us*, 170.

238 **"The long trend is toward a warmer earth":** Carson, *Sea Around Us*, 172.

239 **halfway through its life . . . within about 5 billion years:** "Our Sun: Facts," NASA Science website.

240 **the Capitalocene:** Jason W. Moore, "The Capitalocene, Part I: On the Nature and Origins of Our Ecological Crisis," *The Journal of Peasant Studies* 44, no. 3 (May 4, 2017): 594–630.

Index

Aboriginal people of Australia, 35
Alaska Native Haida, 55
alfalfa, 155–56
algae, 21, 29, 174, 199, 203
American Geosciences Institute, 88
ammonite, 216
amphibians, 208
Andes mountains, 81
animals
  earliest, 96
  evolution of, 136–38
  fossils, 83, 100, 208
  influence of Mesozoic climate on, 227
  influence of plants on, 226–27
  influence on terrestrial environments,
    149–51
  invertebrates, 174–75
  oldest fossils, 100
Antarctica, 95, 102
Anthropocene, 226
Archean eon
  as beginning of rock record, 14
  end of, 45–47
  and manganese, 40–42, 44
  in Minnesota, 49–51, 56–57, 59
Arizona State University, 27
arm metaphor, for Earth timeline, *x,* 15, 71,
  135, 187
ashes, volcanic, 122
asteroids, 209, 231, 231–33
atmosphere; *See also* Great Oxidation event
  in Earth system, 5–6, 30
  moisture in, 178, 198
  and mud, 137
  and volcanoes, 189–90
atmospheric carbon dioxide
  as heat trap, 192

  in Triassic period, 201–2
  and volcanic activity, 150
  and warming oceans, 194
  and weathering of rock, 142, 178
atmospheric oxygen
  in Carboniferous period, 188
  and cyanobacteria, 43–45
  early absence of, 19–20, 26, 37–38
  and fire, 198–200
  and ice ages, 70, 76, 97
  and iron, 50, 62
  and living things, 16
  reactivity of, 27
Australia, 21, 35, 81, 99–100, 103–4, 107–
  14, 121
Avalon Peninsula (Mistaken Point), 120–22,
  127

bacteria, 40, 95, 207; *See also*
  cyanobacteria
Bad Religion (punk band), 139
Bahamas, 21, 81
banded iron formations (BIFs), 55–60, 64
Barghoorn, Elso, 18–20, 25, 28–29, 61,
  67–68
basalts, 56–57, 64, 239
bedrock, striations in, 80, 82, 220
Benson, Roger, 213
Bible, 33–34, 169
BIFs (banded iron formations), 55–60, 64
Bighorn Canyon National Recreation
  Area, Wyo., 1–2
biogeochemistry, 25
biogeomorphology, 152–57
biosphere, 30
Block, Amy Radakovich, 50–51, 64
Bois Forte Band of Ojibwe, 51–54

INDEX

269

Bois Forte Heritage Center and Cultural
  Museum, 53–54
Bolivia, 139–41
Bonsor, Joe, 214, 216, 217, 219–20, 222–24
Botanical Gazette, 201
brachiopods, lingulid, 207
braided streams, 152–53, 155–56
Brengman, Latisha, 51, 63, 64, 65–66
Budyko, Mikhail, 84–85, 105
buns, magmatic, 56

calcretes, 165
California Institute of Technology, 23, 38,
  91
Cambrian period; *See also* Snowball Earth
  theory
  explosion of life, 97–98, 126–31
  fossils, 83, 118
Canada, 18, 36, 193–94
Canadian Shield, 37
carbonates, 81–82, 89–93, 91, 106, 115
carbon cycle, 150, 175–79, 177–78, 199
carbon dioxide, 106, 137, 190, 192, 201–2;
    See also atmospheric carbon dioxide
Carboniferous period, 188
carbon isotopes, 91–93
carbon sinks, 106
Carson, Rachel, v, 15, 236–38
  The Sea Around Us, v, 15, 236
Cenozoic era, 188
charcoal, 196–98
chemical proxies, 13–14
chemostratigraphy, 229
cherts, 18–19; *See also* Gunflint Chert
Chicago, Ill., 178
Children's Museum of Indianapolis, 215,
  218, 226
Chin, Karen, 204–7, 229
  The Clues Are in the Poo, 205
Chinook salmon, 14
chloroplasts, 41
clay
  accumulation, as mud, 135–37, 153
  chemical weathering of, 142–43
  end-Mesozoic layer, 231
  glauconite, 223
  particles, 160–64

physical erosion of, 176
Cloud, Preston, 19, 20
Clues Are in the Poo, The (Chin), 205
coal, 190, 196
Cohen, Phoebe, 96–98
cohesion, of mud particles, 137, 153, 155
Colby College, 195
Colorado College, 24, 102, 118
Columbia University, 87
computer models, 179–81, 203–4
conservation of mass, 178
coprolites, 204–7
copulation, internal, 136
coral, 121, 125, 166
core, Earth's, 12, 45
Cornell University, 139
Crafoord Prize in Geoscience, 67
cranes, 154
creative thinking, 40
Cretaceous period, 231
crust, Earth's, 12, 45–48
Cryogenian glaciation, 100, 104–7, 116,
  123
Cryogenian period, 71, 76, 96, 104–7,
  114–16
cryosphere, 30
cyanobacteria
  and GOE, 25–29, 47–48
  manganese-hoarding, 43–44
  origin of, 21, 68–69

daily routine, in field seasons, 112–14
D'Antonio, Michael, 178–79
Darwin, Charles, 16–19, 32–34, 71, 152,
  156, 174
  On the Origin of Species, 16–17, 180
Davies, Neil, 138–44, 146–49, 156–62,
  160–62, 164, 179–82
debris, organic, 162–63
Deinococcus radiodurans (bacteria), 41
Devonian period, 144, 155, 170, 171, 174,
  175
Devon Island, Canada, 205–6
dichromatic vision, 227
Dick, Robert, 136
Dingle Peninsula, Ireland, 144, 144–46, 167
Dinosaur Provincial Park, Canada, 229

270 INDEX

dinosaurs; *See also* sauropods
  environments of, 225
  extinction of, 231–32
  fossils, 210–11, 215, 219
  fragments of, 222
  and plants, 226
  rise of, 209–12
  and sediments, 218
discs, 122–23, 195–96
diversification of life, 96–98
Dodge, Ossian Euclid (Oro Fino), 52
Dollyphyton boucotii (moss), 160
Drill Core Library, 63
drill cores, 40, 60–66, 64
Dry Valleys of Antarctica, 96, 102
dunes, 102, 145–46
Dylan, Bob, 62–63

earthquakes, 89, 169
Earth system, interconnections in, 30–31;
  See also geobiology
Earth System, The (Kump), 47
Earth timeline, as arm metaphor, *x,* 15, 71,
  135, 187
earthworms, 152
Eberth, David, 229–30
Eckerson, Sophia, 201
ecosystems, rate of change, 230
Ediacaran period, 120–26
Egerton, Victoria, 226, 228
England
  ammonite fossils in, 216
  geologic map of, 35
  millipede fossil in, 180, 188
  turtle fossils in, 203
eons, 188
equatorial regions, 81
erosion, 33, 56, 126, 142, 176, 199
Essig, Jim, 60, 62
Ewing, Ryan, 102, 107–12, 117

fats, fossilized, 203, 206
feldspars, 177
ferricretes, 165–66
Field Museum of Natural History, 178
Fischer, Woody
  early life, 23–24

  graduate students, 26
  and Kaapvaal Craton cores, 41–43
  on manganese-hoarding cyanobacte-
    ria, 44
  on mineral proxies, 37–39
  on mud, 138
  research interests, 160–62
  theory on GOE, 26–29, 47–48
  wedges, 114
fish
  evolution of, 136
  gill-stuffed, 140–41
  origins of, 139
fishing, 12, 18, 124, 237
fjords, 103
Flinders Ranges (Australia), 99–100, 102–3
flocculation, 160–64
floodplains, 137, 175
floodwaters, 137
Floyd, George, 54
food chain, Arctic, 207
fool's gold (pyrite), 38–39, 43, 61, 62, 64
forced relocation, of tribal communities,
  52–54
Forces of Nature (McNeill and Reser), 88
Fortune Head, 128–29
fossil fuels, 190, 193
fossils
  absence in oldest rock layers, 16–19
  ammonite, 216
  Cambrian, 83, 118, 128–29
  corals, 166
  dinosaur, 210–11, 215, 219
  Ediacaran, 121–23, 121–26, 126
  fats, 203, 206
  feces, 204–7
  fish, 139–40
  forests, 173
  in Gunflint Chert, 29, 36
  Jurassic, 221, 224, 226
  leafy, 198–202
  millipede, 147–48, 180, 188
  of Morrison Formation, 211–13, 221
  plankton, 206
  trace, 157, 159
  turtle, 203
fuel combustion, 198

INDEX

Galloway, Jennifer, 200
gases; *See also* carbon dioxide; oxygen
greenhouse, 191–94, 199
of iron, 21–22
ozone-depleting, 190
seafloor, 21
sources and sinks, 46–47
Gaskiers glaciation, 100, 117–20, 119–20, 123
geobiology, 24, 25, 31
geochemistry; *See also* "whiffs of oxygen"
and age of Earth, 35
of carbonates, 92–93, 115
and cyanobacteria, 68–69
of iron, 51, 58, 62, 64–65
proxies, 204
and steranes, 96
geochronologists, 35
geographic isolation, of life forms, 96–97
Geological Survey of Canada, 36, 90, 200
geologic maps, 35
geologic record, incomplete, 180–81
geologic timeline, 32
geology, value of, 181–82
geosciences, Native students in, 54–55
glacial deposits, 79–84
glacial dropstones, 80, 82
glacial ice, 75–76, 81
glacial till, 80, 82–83
glaciations, *See* Cryogenian glaciation; Gaskiers glaciation; Huronian glaciation; Marinoan glaciation; Snowball Earth theory; Sturtian glaciation
glaciers, coastal, 77
Glasspool, Ian, 195–200, 229
glauconite, 223
Global Stratotype Section and Points (GSSPs), 127–28
global transformation, moments of, 24–25
GOE, *See* Great Oxidation Event
GOE proxies, 43, 58
gold deposits, 52–53
golden spikes, 127–28
Goswami, Anjali, 213
Gould, Stephen Jay, 119, 125
Wonderful Life, 127
Graffin, Greg, 139–40

Great Barrier Reef (Australia), 81
Great Dying, 189
Great Infra-Cambrian Ice Age, 84–85, 91; *See also* Snowball Earth theory
Great Oxidation Event (GOE)
and BIFs, 58
and cyanobacteria, 47–48, 68–69
and Woody Fischer, 24–29
onset, 24–27
and radiometric dating, 36
unanswered questions, 65–70
greenhouse gases, 191–94, 199
Greenland, 190, 198, 200, 237
greywacke, 168
GSSPs (Global Stratotype Section and Points), 127–28
guar gum, 163
Gunflint Chert, 18–19, 21, 29, 36, 42, 67

Hadean eon, 14
Hall, James, 167–69
halos, root, 170–71
Harland, Brian, 79–89, 104, 202, 238
Harvard University
Elso Barghoorn, 18
Center for the Environment, 69
Phoebe Cohen, 96
Woody Fischer, 24, 26, 31
Paul Hoffman, 90–93, 99
Linda Ivany, 119
Andy Knoll, 28, 67
Dan Schrag, 94
hearing loss, 61
heat, Earth's early, 45–46
hematite, 65
Henneguya salminicola (salmon parasite), 14
Hibbing, Minn., 62
Hindu texts, 34
His Dark Materials (Pullman), 77
HMS *Beagle,* 34
Hoffman, Paul, 24, 89–95, 97–99, 101–2, 111, 128
Holland, Dick, 24
Hooke, Robert, 202–3
Horner, Jack, 204–5
Hudson Bay, 37
humans, rise of, 232–33

INDEX

Huronian glaciation, 70
Hutton, James, 33–35, 147, 167–70
hydrosphere, 30
hydrothermal vents, 81, 95

ice ages, 69–71, 81–84, 89, 105–7; *See also*
    Snowball Earth theory
Indigenous Geoscience Community
    (IGC), 55
Indigenous peoples, 31, 34, 50–55
Indonesia, 81
Indy (woman at Jurassic Mile), 219
interconnection, 30–31
Intergovernmental Panel on Climate
    Change, 107, 203
internal copulation, 136
International Commission on Stratigra-
    phy, 127
International Union for the Conservation
    of Nature Species, 208
invertebrates, 152, 174–75
Iran, 121
Ireland, 144–45
iron; *See also* banded iron formations;
    ferricretes; Soudan Iron Mine
    (Minnesota)
    dissolved, 49–54, 58
    formations, 64–65
    gases of, 21–22
    geochemistry, 51, 58, 62, 64–65
    industry, 52–54, 62
    magnetic grains of, 104
    rusted layers of, 65
Ivany, Linda, 119, 125, 125–26

jellyfish, 121, 125
Johnson, Aleisha, 48
Journal of Vertebrate Paleontology, 139
Jurassic Mile, 215–17
Jurassic Park (film), 204
Jurassic period, 208, 214, 221, 224, 226

Kaapvaal Craton cores, 40–41, 44
Kalahari Manganese Field, 41
kalpas, 34–35
katabatic winds, 77
Kenrick, Paul, 226–28

Kirschvink, Joe, 91, 106
Knoll, Andy, 25, 28, 67–69, 96
komatiite, 46
krypton, 20
Kump, Lee, 92
    The Earth System, 47

Laguna Pueblo people, 34
Lake Superior, 18
Lake Vermilion (Onamanii-zaaga'igan), 52,
    52–53
Lake Vermilion-Soudan Underground
    Mine State Park (Minnesota), 49, 60
Landing, Ed, 128
leaf litter, 160
leaf shape, 198
lick tests, 220, 221, 225
lightning, 198
Lingappa, Usha, 39–44
lingulid brachiopods, 207
lithosphere, 30
Long, John, 136
Lyell, Charles, 34, 156

MacArthur "genius grant," 69
magma, 56, 88, 188–91
magnetic mineral grains, 104–5
magnetostratigraphy, 229
Maidment, Susie, 210–16, 220–26, 228–30
Maine, 3, 11, 193–95, 239–40
Maloof, Adam, 99–100, 112–13, 115–16
mammals, placental, 232
manganese, 41–45
mantle, Earth's, 12–14, 45–46, 68, 88, 150,
    190
Man Who Found Time, The (Repcheck), 34
marine ecosystem, Mesozoic, 206
marine organisms, 92
Marinoan glaciation, 70–71, 100–101
Mark (guide), 122
mass, conservation of, 178
mass extinctions
    end-Cretaceous, 231
    end-Mesozoic, 209, 231–33
    end-Permian, 188–90, 189–90, 207, 230
    end-Triassic, 195, 198, 200–202
mass spectroscopy, 35

INDEX

mattresses, magmatic, 56
McMahon, Will, 141–43, 148, 157, 160, 175–77
McMurdo Dry Valleys (Antarctica), 96
McNeill, Leila, *Forces of Nature,* 88
McPhee, John, 14
meandering streams, 152–53, 155
Memorial University of Newfoundland, 124, 127–28
Mesozoic climate, influence on plants and animals, 227
Mesozoic era, 188–94, 198–99, 205–17, 229–31, 238–40; *See also specific periods, e.g.:* Jurassic period
Michigan, 51
Microbrachius dicki (fish), 136
microfossils, 27–29, 61, 196
microscopy, 64–65, 157, 176, 196–97
Mid-Atlantic Ridge, 87
millipede fossils, 147–48, 180, 188
Minard Castle (Ireland), 145
mineral exploration, 17
mineral proxies, 13–14
minerals, dissolving in oxygen, 19
Minnesota, 24, 49–51, 53, 58, 118
Minnesota Department of Natural Resources, 63
Minnesota Geological Survey, 50, 54–55
Missouri River, 154
Mistaken Point (Avalon Peninsula), 120–22, 127
Mistaken Point Ecological Reserve, 122
Mitchell, Sherri, *Sacred Instructions,* 31
Monthly Review, 34
Morrison Formation, 211, 211–14, 221–26
mud
    clay accumulation, 135–37, 153
    and invertebrates, 174–75
    on land, 160–62, 174–75, 179
    particle cohesion, 137, 153, 155
    and plants, 150–53, 170–71, 175
    rise of, 135–39, 143, 161, 179, 188
    and rivers, 136–37
mudflows, 82–83
mudrock, 141
mudstones, 164, 166, 176–77, 206
multicellular life, 71, 94–98, 124, 238

mundane, appreciation of the, 146–49, 147–48, 176
Myrow, Paul, 117–20, 126, 127–28, 128–29

Namibia, 90–91, 93, 102, 111, 121
Narbonne, Guy, 128
narrow-leaved plants, 198
National Association for Interpretation, 54
National Science Foundation, 76
National Trust for Historic Preservation, 54
Native Americans, 31, 34, 50–55
Natural History Museum (London), 210, 214, 224, 226
Naturalis Biodiversity Center (Netherlands), 216
Nebraska, 154
nematophytes, 197
neon, 20
New England region, 194, 195
Newfoundland, 117–31
New York State Geological Survey, 128
New York Times, 236
Nhat Hanh, Thich, *No Mud, No Lotus,* 138
nitrogen, 192
No Mud, No Lotus (Nhat Hanh), 138
Northwest Territories, 36
Norway, 82
Nova Scotia, 193
Nunavut, Canada, 205
nutrients, outpouring from roots, 173–75
Ny-Ålesund, Norway, 77–79

Oak Ridge National Laboratory, 200
Obama, Barack, 69
oceans, 96, 194
oil, 190, 193
Ojibwe people, 51–52
Oliver, Mary, 138
Onamanii-zaaga'igan (Lake Vermilion), 52
On the Origin of Species (Darwin), 16–17, 180
ooids, 65
optical analysis, 64
ordinary, appreciation of the, 146–49
organic debris, 162–63
Oro Fino (Dodge, Ossian Euclid), 52
Ossian Euclid Dodge, 52

INDEX

oxygen; *see also* atmospheric oxygen; cyanobacteria; GOE (Great Oxidation Event); "whiffs of oxygen"
  earliest producers, 20–21
  early absence of, 36–39
  inconsistency of concentration, 66
  lack of insulating power, 192
  minerals dissolving in, 19
  origin of, 16–20
  from plants, 136
  and rise of complex life forms, 97–98
  sinks, 46–47
ozone-depleting gases, 190

paleoclimate proxies, 200–208
Paleozoic era, 188; *See also* Cambrian period; Carboniferous period; Devonian period; Silurian period
Palisades, 191
Pangea, 188–89, 209, 239
Paola, Chris, 154–55
Parton, Dolly, 160
pebbles, self-organization of, 158–59
Pennsylvania State University, 47
Penobscot Nation, 31
periods, geologic, 188
Pernatty Lagoon, 107
persistence of life forms, during Cryogenian period, 96
phosphorus, 68–69, 199
photosynthesis, 20–22, 40–41, 44–45, 68, 92
phyllosilicates, 161
physical proxies, 13–14
pillows, magmatic, 56
"pizza discs," 122–23
placental mammals, 232
plankton, fossilized, 206
plants
  and dinosaurs, 226
plant(s)
  and chemical weathering, 176–77
  distribution, 227
  influence of Mesozoic climate on, 227
  influence on animals, 226–27
  influence on rivers, 153–56

  influence on terrestrial environment, 149–51, 173–75
  and mud, 150–53, 170–71, 175
  and mudrock, 141–42, 175–76
  rise of, 136, 188
  shaping of environment, 226–29
plate tectonics, 56, 59, 61, 86–90, 106
Platte River, 154
Platte River Recovery Implementation Program, 154, 156
Playfair, John, 168–71
polar regions, 84, 110
Poppick, Laura, 3–4, 113, 235–36
porcelain shards, 11–12
Precambrian-Cambrian golden spike, 126–31
President's Council of Advisors for Science and Technology, 69
primordial warmth, 45
Princeton University, 99
Principles of Geology (Lyell), 34, 156
Proterozoic eon, 46
Proulx, Annie, *The Shipping News,* 117
proxies
  geologic, 13–14
  for GOE, 43, 58
  for oxygen, 37–39, 98
  for paleoclimate, 200–208
Pullman, Philip, *His Dark Materials,* 77
pyrite, 38–39, 43, 61, 62, 64

quartz, 52, 59
Quebec, 193
Queen's University, 128

radioactive decay, 45
radiometric dating, 35
red sandstone, 169–70
Repcheck, Jack, 169
  The Man Who Found Time, 34
Reser, Anna, *Forces of Nature,* 88
resin discs, 195–96
"Rice" (Oliver), 138
rift valleys, 88, 89
riverbanks, 136
rivers
  deposits, 157–60

INDEX

influence of plants on, 153–56
and mud, 136–37
shapes of, 152–56
rock
descriptions, 223–24
influence of roots on, 164–66
outcrops, 167–68
weathering of, 142–43, 176–77
roots
influence on rock, 164–66
outpouring of nutrients, 173–75
remnants, 170–72
water flow through, 150
weathering via, 142
Roscoe, Stu, 36–37
Rose, Catherine, 114
Royal Society of Edinburgh, 33
Rudwick, Martin, 83–84
Russia, 121

Sacred Instructions (Mitchell), 31
St. Anthony Falls Laboratory (Minneapolis, Minn.), 118, 155
St. Paul Pioneer, 52
salmon, 14
sandstone, 169–70
sand wedges, 107–10, 114
Santos, Fernanda, 200
sauropods, 218–22, 226–27
Savoy, Lauret, *Trace,* 54
schistus, 168
Schoene, Blair, 112, 117–18
Schopf, William, 19
Schrag, Dan, 69–70, 93, 97
Science (journal), 19, 93, 163
Scientific American, 83
Scotland, 33
Sea Around Us, The (Carson), v, 15, 236
seafloor
becoming a mountaintop, 168–69
and BIFs, 57–59
gases, 21, 21–22
and hydrothermal vents, 81, 95
and iron, 49–50
and manganese, 42, 44
mapping, 77, 87, 89

and plate tectonics, 87
sediments, 18, 20, 26, 168, 193, 204
and volcanic ash, 122
and volcanic rock, 56
sea level change, 66
seawater
and carbon isotopes, 92
fast-warming, 194
freezing of, 187
meeting fresh water, 76
oxygenated, 65
temperature, 13, 56, 77, 232
sedimentary rocks
interpretation of, 12–14, 180
and roots, 164–65
stratigraphic logs of, 213
sediment(s)
and dinosaurs, 218
in fjords, 103
and floodplains, 137
and GOE, 25–26
Hutton's observation of, 33
and invertebrates, 175
in Ny-Ålesund, 77
in rivers, 153
in seafloors, 18, 20, 26, 168, 193, 204
sizes of, 135
unsorted mixes of, 80
uranium in, 37
"Sex That Moves Mountains" (Fremier), 151
Shillito, Anthony, 174
Shipping News, The (Proulx), 117
Siberia, 188
Siccar Point, Ireland, 168–70
silica, 65
Silko, Leslie Marmon, 34
Silurian period, 144, 155, 170, 175, 197, 239
silver deposits, 52
Sixth Assessment Report (Intergovernmental Panel on Climate Change), 107, 203
Slea Head (Ireland), 157, 176
Slushball Earth, 99–104, 101, 107, 109, 114
Smith College, 201

INDEX

Snowball Earth theory, 91–102, 106–11, 115–16; *See also* Slushball Earth
solar constant, 85
Soudan Iron Mine (Minnesota), 49, 51–52, 63
South Africa, 39
Soviet Union, 84
speciation, 96–97
spiders, 103–4
sponges, 96
steranes, 96
Steno, Nicolas, 17
stomata, 200, 202
storm activity, 66
stratigraphic logs, 99
stratigraphic record, incompleteness of, 180–81
streams, 152–56
striations, in bedrock, 80, 82
stromatolites, 21, 42–43, 45
Strong, Jaylen, 53
Sturtian glaciation, 70–71, 100, 100–101, 106
sulfur, 21–22, 26
sun, 239
Sundance Sea, 218, 222
Svalbard, Norway, 76–77, 76–79, 79–80, 82, 141, 202, 237
Sweden, 67
Syracuse University, 118–19

Tal, Michal, 154–56
Tarhan, Lidya, 175
tectonic plates, *See* plate tectonics
terrestrial environment(s)
    evolution of, 141–42
    influence of animals on, 149–51
    influence of plants on, 149–51, 173–75
Tharp, Marie, 87–89
thunderstorms, 198
Tierney, Jessica, 107, 203, 206
till, glacial, 80, 82–83
tillites, 119
Todd, Wendy F. K'ah Skaahluwaa, 55
Trace (Savoy), 54
treaties, US government and Bois Forte Band of Ojibwe, 52–54

tree canopies, 160
Treptichnus pedum, 126–31, 128–29
Triassic period, 189–90; *See also* mass extinctions/end-Triassic
tribal consent, 54
trichromatic vision, 227
turtles, fossilized, 203
Tyler, Stanley, 17–19, 28–29, 32, 42, 61, 67–68

UNESCO World Heritage Sites, 121–22
unfrozen habitats, 95–96
University of Arizona, 47–48, 107, 203
University of California Berkeley, 39
University of Cambridge, 79, 126, 138, 213
University of Colorado Boulder, 204
University of Johannesburg, 40
University of Manchester, 215
University of Minnesota, 154–55
University of Minnesota Duluth, 51, 55
University of Oxford, 213
University of Saskatchewan, 174
University of Victoria, 90
uranium, 17, 36–38, 43–45, 61, 212, 238
U.S. Steel Corporation, 49, 63

Veenma, Yorick, 149–50, 157, 164, 166, 171–72, 181
vision, evolution of, 227
visual arts, 40
Vladimir Chimp and the Space Barnacles, 24
volcanoes (volcanism); *See also* magma
    ash, 122
    basalts, 56
    and carbon dioxide, 150
    and end-Permian mass extinction, 188–90
    and oxygen, 20
    and plate tectonics, 86, 90, 106
    underwater, 46–47
    and wildfires, 195, 198

Wabanaki Nations, 12
Wales, 35, 196
Waterbelt, 101
weather, record-breaking, 193–94

INDEX

weathering, 142–43, 157, 176–79
Wegener, Alfred, 88
"whiffs of oxygen"
    Woody Fischer on, 27–28, 101
    and iron, 51, 62, 64–66
    timing of, 45–48
White Earth Disaster, 85, 105
whooping cranes, 154
wildfires, 136, 193–200, 202, 230
Williams College, 96
Wine Strand, 170
Wisconsin, 17, 51, 214–19
women in sciences, 88–89
Wonderful Life (Gould), 127

Woodward, Ian, 202
World Ocean Floor Map, 89
World War II, 79
Wyoming, 1–2

xanthan gum, 163
xenon, 20
X-ray fluorescence, 43

Yale University, 175
Yucatan Peninsula, 231, 231–33

Zeichner, Sarah, 161–62
zircons, 36, 44–45, 122, 212, 229